Timelines of Discovery in Physics

" Classical to Quantum "

Edited by Paul F. Kisak

Contents

Chapter 1

Chronology of the universe

See also: Timeline of the formation of the Universe

The **chronology of the universe** describes the history and future of the universe according to Big Bang cosmology, the

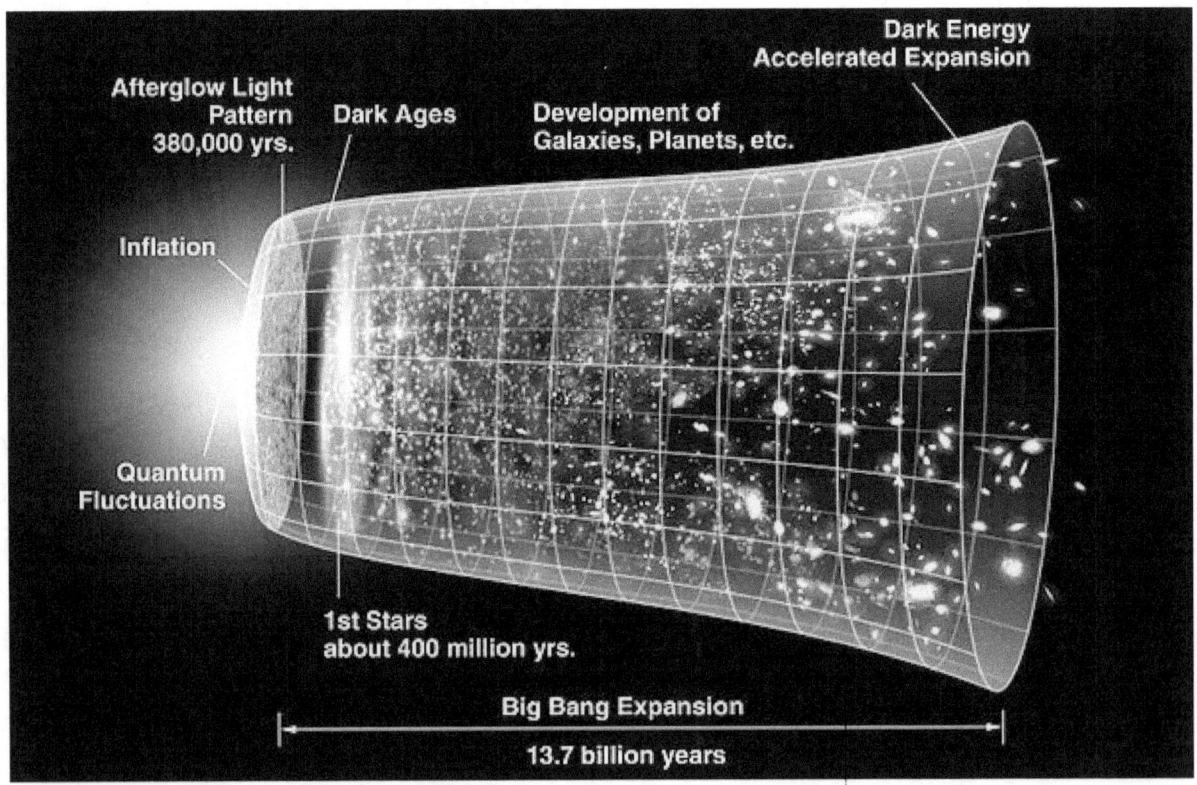

Diagram of Evolution of the universe from the Big Bang (left) - to the present.

prevailing scientific model of how the universe developed over time from the Planck epoch, using the cosmological time parameter of comoving coordinates. The model of the universe's expansion is known as the Big Bang. As of 2015, this expansion is estimated to have begun 13.799 ± 0.021 billion years ago.[1] It is convenient to divide the evolution of the universe so far into three phases.

1.1   Summary

In the first phase, the very earliest universe was so hot, or energetic, that initially no matter particles existed or could exist perhaps only fleetingly. According to prevailing scientific theories, at this time the distinct forces we see around us today were joined in one unified force. Space-time itself expanded during an inflationary epoch due to the immensity of the energies involved. Gradually the immense energies cooled – still to a temperature inconceivably hot compared to any we see around us now, but sufficiently to allow forces to gradually undergo symmetry breaking, a kind of repeated condensation from one status quo to another, leading finally to the separation of the strong force from the electroweak force and the first particles.

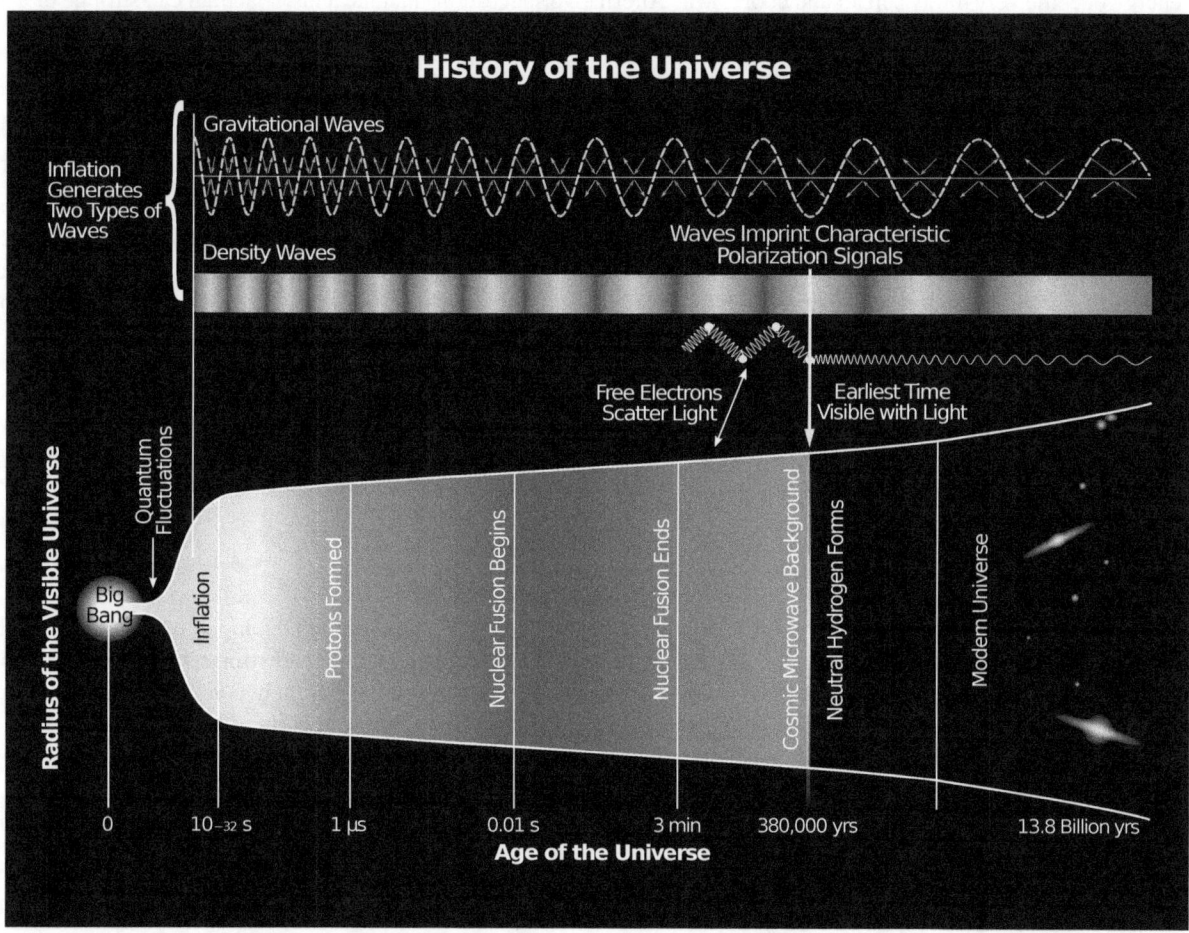

History of the Universe - gravitational waves are hypothesized to arise from cosmic inflation, a faster-than-light expansion just after the Big Bang (17 March 2014).[2][3][4]

In the second phase, the resulting quark–gluon plasma universe then cooled further, the current fundamental forces we know take their present forms through further symmetry breaking – notably the breaking of electroweak symmetry – and the full range of complex and composite particles we see around us today became possible, leading to a gravitationally dominated universe, the first neutral atoms (~ 80% hydrogen), and the cosmic microwave background radiation we can detect today. Modern high energy particle physics theories are satisfactory at these energy levels, and so physicists believe they have a good understanding of this and subsequent development of the fundamental universe around us. Because of these changes, space had also become largely transparent to light and other electromagnetic energy, rather than "foggy", by the end of this phase.

The third phase started after a short dark age with a universe whose fundamental particles and forces were as we know them, and witnessed the emergence of large scale stable structures, such as the earliest stars, quasars, galaxies, clusters of galaxies and superclusters, and the development of these to create the kind of universe we see today. Some researchers call the development of all this physical structure over billions of years "cosmic evolution". Other, more interdisciplinary, researchers refer to "cosmic evolution" as the entire scenario of growing complexity from big bang to humankind, thereby incorporating biology and culture into a unified view of all complex systems in the universe to date.[5]

Beyond the present day, scientists anticipate that the Earth will cease to be able to support life in about a billion years, and will be enveloped by a greatly-expanded Sun in about 5 billion years. On a far longer timescale, the Stelliferous Era will end as stars eventually die and fewer are born to replace them, leading to a darkening universe. Various theories suggest a number of subsequent possibilities. If particles such as protons are unstable then eventually matter may evaporate into low level energy in a kind of entropy related heat death. Alternatively the universe may collapse in a big crunch, although current data shows the rate of expansion is still increasing. If this is correct then it may end in a "big freeze" as matter and energy become very thinly spread and cool down. Alternative suggestions include a false vacuum catastrophe or a Big Rip as possible ends to the universe.

1.2 Very early universe

All ideas concerning the very early universe (cosmogony) are speculative. No accelerator experiments have yet probed energies of sufficient magnitude to provide any experimental insight into the behavior of matter at the energy levels that prevailed during this period. Proposed scenarios differ radically. Some examples are the Hartle–Hawking initial state, string landscape, brane inflation, string gas cosmology, and the ekpyrotic universe. Some of these are mutually compatible, while others are not.

1.2.1 Planck epoch

0 to 10^{-43} second after the Big Bang

Main article: Planck epoch

The Planck epoch is an era in traditional (non-inflationary) big bang cosmology wherein the temperature was so high that the four fundamental forces—electromagnetism, gravitation, weak nuclear interaction, and strong nuclear interaction—were one fundamental force. Little is understood about physics at this temperature; different hypotheses propose different scenarios. Traditional big bang cosmology predicts a gravitational singularity before this time, but this theory relies on general relativity and is expected to break down due to quantum effects.

In inflationary cosmology, times before the end of inflation (roughly 10^{-32} second after the Big Bang) do not follow the traditional big bang timeline.

1.2.2 Grand unification epoch

Between 10^{-43} second and 10^{-36} second after the Big Bang[6]

Main article: Grand unification epoch

As the universe expanded and cooled, it crossed transition temperatures at which forces separate from each other. These are phase transitions much like condensation and freezing. The grand unification epoch began when gravitation separated from the other forces of nature, which are collectively known as gauge forces. The non-gravitational physics in this epoch would be described by a so-called grand unified theory (GUT). The grand unification epoch ended when the GUT forces further separate into the strong and electroweak forces.

1.2.3 Electroweak epoch

Between 10^{-36} second (or the end of inflation) and 10^{-32} second after the Big Bang[6]

Main article: Electroweak epoch

According to traditional big bang cosmology, the Electroweak epoch began 10^{-36} second after the Big Bang, when the temperature of the universe was low enough (10^{28} K) to separate the strong force from the electroweak force (the name for the unified forces of electromagnetism and the weak interaction). In inflationary cosmology, the electroweak epoch ends when the inflationary epoch begins, at roughly 10^{-32} second.

Inflationary epoch

Unknown duration, ending 10^{-32}(?) second after the Big Bang

Main article: Inflationary epoch

Cosmic inflation was an era of accelerating expansion produced by a hypothesized field called the inflaton, which would have properties similar to the Higgs field and dark energy. While decelerating expansion would magnify deviations from homogeneity, making the universe more chaotic, accelerating expansion would make the universe more homogeneous. A sufficiently long period of inflationary expansion in the past could explain the high degree of homogeneity that is observed in the universe today at large scales, even if the state of the universe before inflation was highly disordered.

Inflation ended when the inflaton field decayed into ordinary particles in a process called "reheating", at which point ordinary Big Bang expansion began. The time of reheating is usually quoted as a time "after the Big Bang". This refers to the time that would have passed in traditional (non-inflationary) cosmology between the Big Bang singularity and the universe dropping to the same temperature that was produced by reheating, even though, in inflationary cosmology, the traditional Big Bang did not occur.

According to the simplest inflationary models, inflation ended at a temperature corresponding to roughly 10^{-32} second after the Big Bang. As explained above, this does not imply that the inflationary era lasted less than 10^{-32} second. In fact, in order to explain the observed homogeneity of the universe, the duration must be longer than 10^{-32} second. In inflationary cosmology, the earliest meaningful time "after the Big Bang" is the time of the end of inflation.

On March 17, 2014, astrophysicists of the BICEP2 collaboration announced the detection of inflationary gravitational waves in the B-mode power spectrum which was interpreted as clear experimental evidence for the theory of inflation.[2][3][4][7][8][9] However, on June 19, 2014, lowered confidence in confirming the cosmic inflation findings was reported [8][10][11] and finally, on February 2, 2015, a joint analysis of data from BICEP2/Keck and Planck satellite concluded that the statistical "significance [of the data] is too low to be interpreted as a detection of primordial B-modes" and can be attributed mainly to polarized dust in the Milky Way.[12][13][14][15]

Baryogenesis

Main article: Baryogenesis

There is currently insufficient observational evidence to explain why the universe contains far more baryons than antibaryons. A candidate explanation for this phenomenon must allow the Sakharov conditions to be satisfied at some time after the end of cosmological inflation. While particle physics suggests asymmetries under which these conditions are met, these asymmetries are too small empirically to account for the observed baryon-antibaryon asymmetry of the universe.

1.3   Early universe

After cosmic inflation ends, the universe is filled with a quark–gluon plasma. From this point onwards the physics of the early universe is better understood, and less speculative.

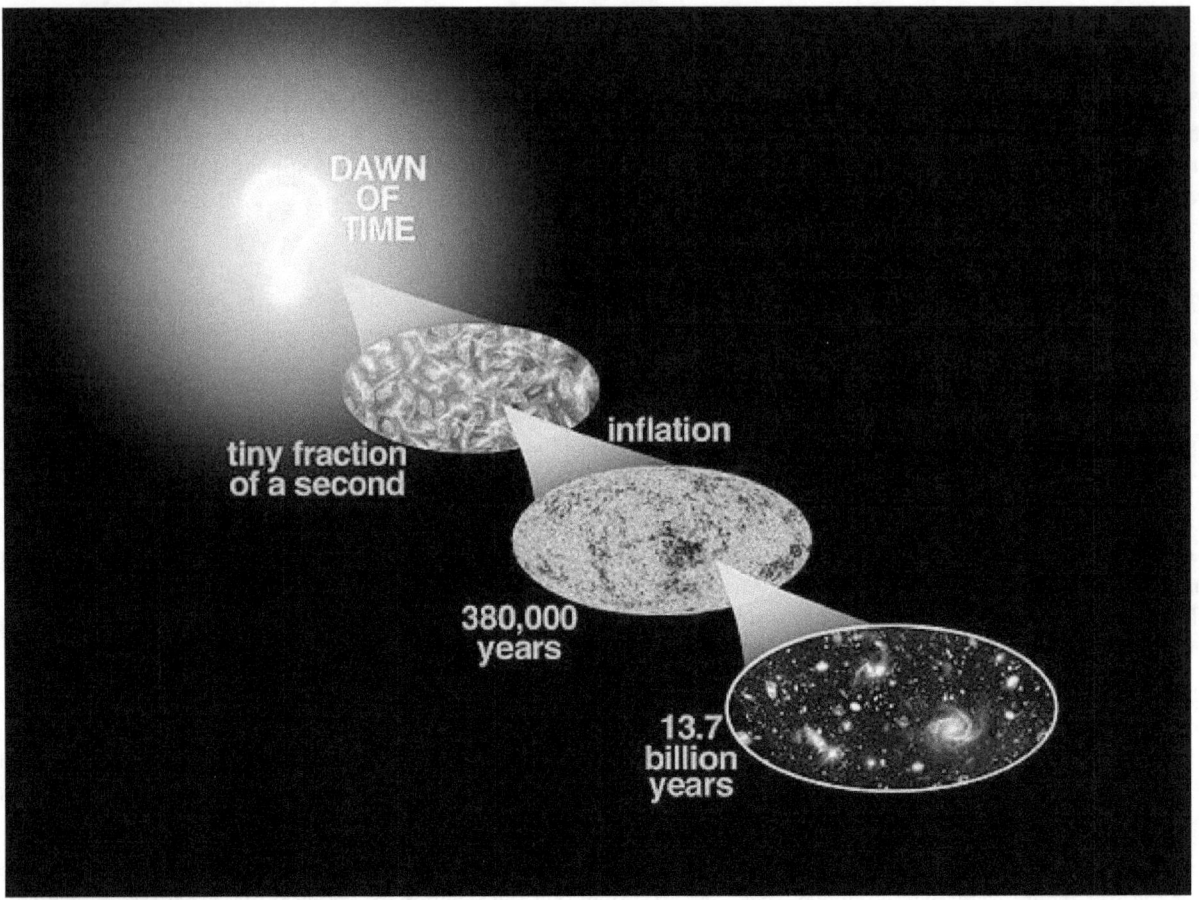

Cosmic History

1.3.1 Supersymmetry breaking (speculative)

Main article: Supersymmetry breaking

If supersymmetry is a property of our universe, then it must be broken at an energy that is no lower than 1 TeV, the electroweak symmetry scale. The masses of particles and their superpartners would then no longer be equal, which could explain why no superpartners of known particles have ever been observed.

1.3.2 Electroweak symmetry breaking and the quark epoch

Between 10^{-12} second and 10^{-6} second after the Big Bang

Main articles: Electroweak symmetry breaking and Quark epoch

As the universe's temperature falls below a certain very high energy level, it is believed that the Higgs field spontaneously acquires a vacuum expectation value, which breaks electroweak gauge symmetry. This has two related effects:

1. The weak force and electromagnetic force, and their respective bosons (the W and Z bosons and photon) manifest differently in the present universe, with different ranges;

2. Via the Higgs mechanism, all elementary particles interacting with the Higgs field become massive, having been massless at higher energy levels.

At the end of this epoch, the fundamental interactions of gravitation, electromagnetism, the strong interaction and the weak interaction have now taken their present forms, and fundamental particles have mass, but the temperature of the universe is still too high to allow quarks to bind together to form hadrons.

1.3.3 Hadron epoch

> *Between 10^{-6} second and 1 second after the Big Bang*

Main article: Hadron epoch

The quark–gluon plasma that composes the universe cools until hadrons, including baryons such as protons and neutrons, can form. At approximately 1 second after the Big Bang neutrinos decouple and begin traveling freely through space. This cosmic neutrino background, while unlikely to ever be observed in detail since the neutrino energies are very low, is analogous to the cosmic microwave background that was emitted much later. (See above regarding the quark–gluon plasma, under the String Theory epoch.) However, there is strong indirect evidence that the cosmic neutrino background exists, both from Big Bang nucleosynthesis predictions of the helium abundance, and from anisotropies in the cosmic microwave background

1.3.4 Lepton epoch

> *Between 1 second and 10 seconds after the Big Bang*

Main article: Lepton epoch

The majority of hadrons and anti-hadrons annihilate each other at the end of the hadron epoch, leaving leptons and anti-leptons dominating the mass of the universe. Approximately 10 seconds after the Big Bang the temperature of the universe falls to the point at which new lepton/anti-lepton pairs are no longer created and most leptons and anti-leptons are eliminated in annihilation reactions, leaving a small residue of leptons.[16]

1.3.5 Photon epoch

> *Between 10 seconds and 380,000 years after the Big Bang*

Main article: Photon epoch

After most leptons and anti-leptons are annihilated at the end of the lepton epoch the energy of the universe is dominated by photons. These photons are still interacting frequently with charged protons, electrons and (eventually) nuclei, and continue to do so for the next 380,000 years.

Nucleosynthesis

> *Between 3 minutes and 20 minutes after the Big Bang*[17]

Main article: Big Bang nucleosynthesis

During the photon epoch the temperature of the universe falls to the point where atomic nuclei can begin to form. Protons (hydrogen ions) and neutrons begin to combine into atomic nuclei in the process of nuclear fusion. Free neutrons combine with protons to form deuterium. Deuterium rapidly fuses into helium-4. Nucleosynthesis only lasts for about seventeen minutes, since the temperature and density of the universe has fallen to the point where nuclear fusion cannot continue. By this time, all neutrons have been incorporated into helium nuclei. This leaves about three times more hydrogen than helium-4 (by mass) and only trace quantities of other light nuclei.

Matter domination

70,000 years after the Big Bang

At this time, the densities of non-relativistic matter (atomic nuclei) and relativistic radiation (photons) are equal. The Jeans length, which determines the smallest structures that can form (due to competition between gravitational attraction and pressure effects), begins to fall and perturbations, instead of being wiped out by free-streaming radiation, can begin to grow in amplitude.

According to ΛCDM, at this stage, cold dark matter dominates, paving the way for gravitational collapse to amplify the tiny inhomogeneities left by cosmic inflation, making dense regions denser and rarefied regions more rarefied. However, because present theories as to the nature of dark matter are inconclusive, there is as yet no consensus as to its origin at earlier times, as currently exist for baryonic matter.

Recombination

ca. 377,000 years after the Big Bang

Main article: Recombination (cosmology)
Hydrogen and helium *atoms* begin to form as the density of the universe falls. This is thought to have occurred about

9 year WMAP data (2012) shows the cosmic microwave background radiation variations throughout the universe from our perspective, though the actual variations are much smoother than the diagram suggests.[18][19]

377,000 years after the Big Bang.[20] Hydrogen and helium are at the beginning ionized, i.e., no electrons are bound to the nuclei, which (containing positively charged protons) are therefore electrically charged (+1 and +2 respectively). As the universe cools down, the electrons get captured by the ions, forming electrically neutral atoms. This process is relatively fast (and faster for the helium than for the hydrogen), and is known as recombination.[21] At the end of recombination, most of the protons in the universe are bound up in neutral atoms. Therefore, the photons' mean free path becomes effectively infinite and the photons can now travel freely (see Thomson scattering): the universe has become transparent. This cosmic event is usually referred to as *decoupling*.

The photons present at the time of decoupling are the same photons that we see in the cosmic microwave background (CMB) radiation, after being greatly cooled by the expansion of the universe. Around the same time, existing pressure waves within the electron-baryon plasma — known as baryon acoustic oscillations — became embedded in the distribution of matter as it condensed, giving rise to a very slight preference in distribution of large scale objects. Therefore the cosmic microwave background is a picture of the universe at the end of this epoch including the tiny fluctuations generated during inflation (see diagram), and the spread of objects such as galaxies in the universe is an indication of the scale and size of the universe as it developed over time.[22]

Habitable epoch

See also: Abiogenesis

The chemistry of life may have begun shortly after the Big Bang, 13.8 billion years ago, during a habitable epoch when the Universe was only 10-17 million years old.[23][24][25]

Dark Ages

See also: Hydrogen line

Before decoupling occurred, most of the photons in the universe were interacting with electrons and protons in the photon–baryon fluid. The universe was opaque or "foggy" as a result. There was light but not light we can now observe through telescopes. The baryonic matter in the universe consisted of ionized plasma, and it only became neutral when it gained free electrons during "recombination", thereby releasing the photons creating the CMB. When the photons were released (or decoupled) the universe became transparent. At this point the only radiation emitted was the 21 cm spin line of neutral hydrogen. There is currently an observational effort underway to detect this faint radiation, as it is in principle an even more powerful tool than the cosmic microwave background for studying the early universe. The Dark Ages are currently thought to have lasted between 150 million to 800 million years after the Big Bang. The October 2010 discovery of UDFy-38135539, the first observed galaxy to have existed during the following reionization epoch, gives us a window into these times. The galaxy earliest in this period observed and thus also the most distant galaxy ever observed is currently on the record of Leiden University's Richard J. Bouwens and Garth D. Illingsworth from UC Observatories/Lick Observatory. They found the galaxy UDFj-39546284 to be at a time some 480 million years after the Big Bang or about halfway through the Cosmic Dark Ages at a distance of about 13.2 billion light-years. More recently, the UDFj-39546284 galaxy was found to be around "380 million years" after the Big Bang and at a distance of 13.37 billion light-years.[26]

1.4   Structure formation

See also: Large-scale structure of the cosmos and Structure formation
Structure formation in the big bang model proceeds hierarchically, with smaller structures forming before larger ones. The first structures to form are quasars, which are thought to be bright, early active galaxies, and population III stars. Before this epoch, the evolution of the universe could be understood through linear cosmological perturbation theory: that is, all structures could be understood as small deviations from a perfect homogeneous universe. This is computationally relatively easy to study. At this point non-linear structures begin to form, and the computational problem becomes much more difficult, involving, for example, N-body simulations with billions of particles.

1.4.1   Reionization

150 million to 1 billion years after the Big Bang

See also: Reionization and 21 centimeter radiation

The first stars and quasars form from gravitational collapse. The intense radiation they emit reionizes the surrounding universe. From this point on, most of the universe is composed of plasma.

The Hubble Ultra Deep Fields often showcase galaxies from an ancient era that tell us what the early Stelliferous Age was like.

1.4.2 Formation of stars

See also: Star formation

The first stars, most likely Population III stars, form and start the process of turning the light elements that were formed in the Big Bang (hydrogen, helium and lithium) into heavier elements. However, as yet there have been no observed Population III stars, and understanding of them is currently based on computational models of their formation and evolution. Fortunately observations of the Cosmic Microwave Background radiation can be used to date when star formation began in earnest. Analysis of such observations made by the European Space Agency's Planck telescope, as reported by BBC News in early February, 2015, concludes that the first generation of stars lit up 560 million years after the Big Bang. [27] [28]

Another Hubble image shows an infant galaxy forming nearby, which means this happened very recently on the cosmological timescale. This shows that new galaxy formation in the universe is still occurring.

1.4.3   Formation of galaxies

See also: Galaxy formation and evolution

Large volumes of matter collapse to form a galaxy. Population II stars are formed early on in this process, with Population I stars formed later.

Johannes Schedler's project has identified a quasar CFHQS 1641+3755 at 12.7 billion light-years away,[29] when the universe was just 7% of its present age.

On July 11, 2007, using the 10-metre Keck II telescope on Mauna Kea, Richard Ellis of the California Institute of Technology at Pasadena and his team found six star forming galaxies about 13.2 billion light years away and therefore created when the universe was only 500 million years old.[30] Only about 10 of these extremely early objects are currently known.[31] More recent observations have shown these ages to be shorter than previously indicated. The most distant galaxy observed as of October 2013 has been reported to be 13.1 billion light years away.[32]

The Hubble Ultra Deep Field shows a number of small galaxies merging to form larger ones, at 13 billion light years, when the universe was only 5% its current age.[33] This age estimate is now believed to be slightly shorter.[32]

Based upon the emerging science of nucleocosmochronology, the Galactic thin disk of the Milky Way is estimated to have been formed 8.8 ± 1.7 billion years ago.[34]

1.4.4 Formation of groups, clusters and superclusters

See also: Large-scale structure of the cosmos

Gravitational attraction pulls galaxies towards each other to form groups, clusters and superclusters.

1.4.5 Formation of the Solar System

9 billion years after the Big Bang

Main article: Formation and evolution of the Solar System

The Solar System began forming about 4.6 billion years ago, or about 9 billion years after the Big Bang. A fragment of a molecular cloud made mostly of hydrogen and traces of other elements began to collapse, forming a large sphere in the center which would become the Sun, as well as a surrounding disk. The surrounding accretion disk would coalesce into a multitude of smaller objects that would become planets, asteroids, and comets. The Sun is a late-generation star, and the Solar System incorporates matter created by previous generations of stars.

1.4.6 Today

13.8 billion years after the Big Bang

The Big Bang is estimated to have occurred about 13.8 billion years ago.[35] Since the expansion of the universe appears to be accelerating, its large-scale structure is likely to be the largest structure that will ever form in the universe. The present accelerated expansion prevents any more inflationary structures entering the horizon and prevents new gravitationally bound structures from forming.

1.5 Ultimate fate of the universe

Main article: Ultimate fate of the universe

As with interpretations of what happened in the very early universe, advances in fundamental physics are required before it will be possible to know the ultimate fate of the universe with any certainty. Below are some of the main possibilities.

1.5.1 Fate of the Solar System: 1 to 5 billion years

Main articles: Formation and evolution of the Solar System § Future, Stability of the Solar System, Future of the Earth § Solar evolution and Red giant § The Sun as a red giant

Over a timescale of a billion years or more, the Earth and Solar System are unstable. Earth's existing biosphere is expected to vanish in about a billion years, as the Sun's heat production gradually increases to the point that liquid water and life are unlikely;[36] the Earth's magnetic fields, axial tilt and atmosphere are subject to long-term change; and the Solar System

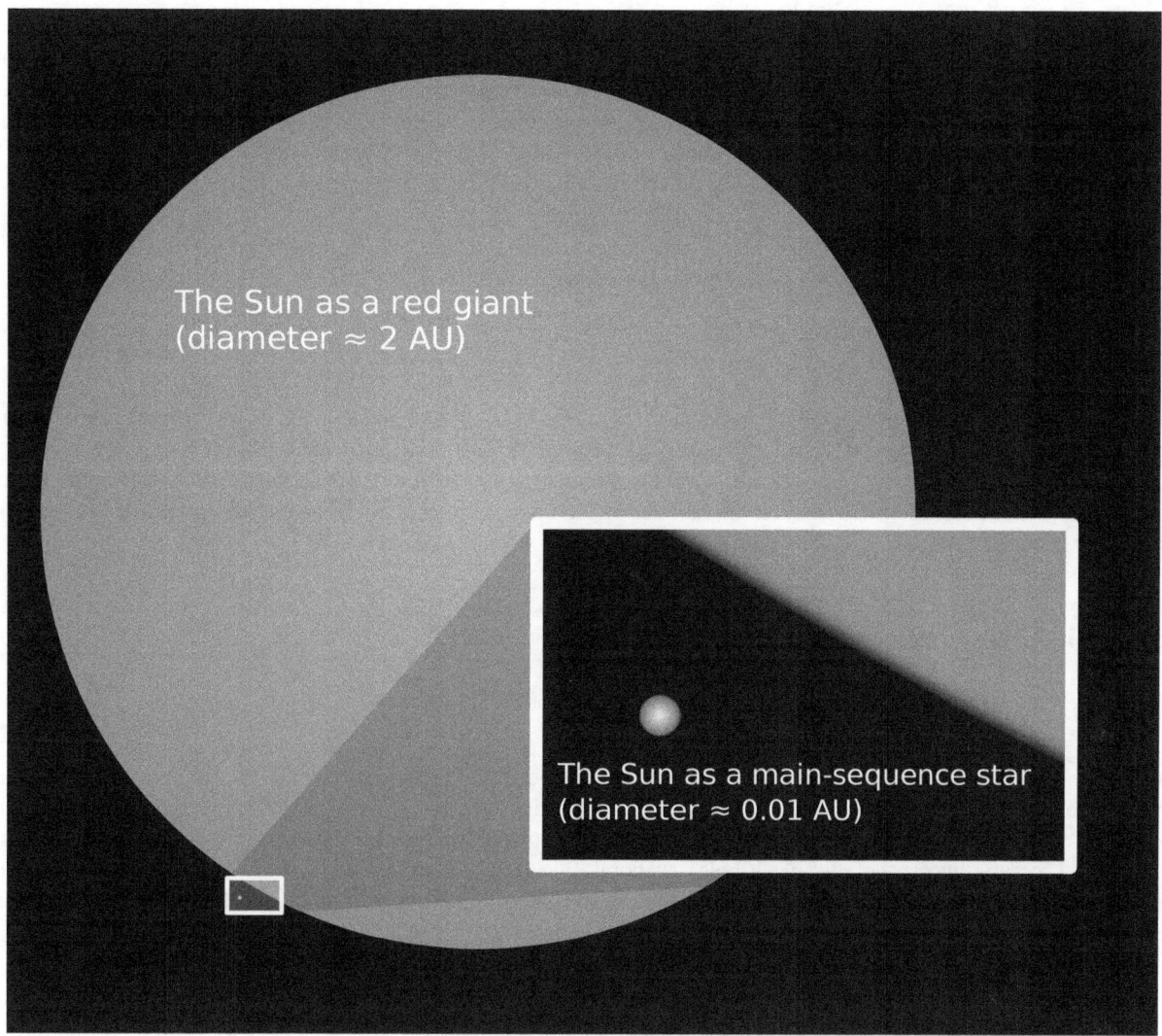

Relative size of our Sun as it is now (inset) compared to its estimated future size as a red giant

itself is chaotic over million- and billion-year timescales.[37] Eventually in around 5.4 billion years from now, the core of the Sun will become hot enough to trigger hydrogen fusion in its surrounding shell.[36] This will cause the outer layers of the star to expand greatly, and the star will enter a phase of its life in which it is called a red giant.[38][39] Within 7.5 billion years, the Sun will have expanded to a radius of 1.2 AU—256 times its current size, and studies announced in 2008 show that due to tidal interaction between Sun and Earth, Earth would actually fall back into a lower orbit, and get engulfed and incorporated inside the Sun before the Sun reaches its largest size, despite the Sun losing about 38% of its mass.[40] The Sun itself will continue to exist for many billions of years, passing through a number of phases, and eventually ending up as a long-lived white dwarf. Eventually, after billions more years, the Sun will finally cease to shine altogether, becoming a black dwarf.[41]

1.5.2   Big Rip: ≥20 billion years from now

See also: Big Rip

This scenario is possible only if the energy density of dark energy actually increases without limit over time. Such dark energy is called phantom energy and is unlike any known kind of energy. In this case, the expansion rate of the universe will increase without limit. Gravitationally bound systems, such as clusters of galaxies, galaxies, and ultimately the Solar System will be torn apart. Eventually the expansion will be so rapid as to overcome the electromagnetic forces holding molecules and atoms together. Finally even atomic nuclei will be torn apart and the universe as we know it will end in an unusual kind of gravitational singularity. At the time of this singularity, the expansion rate of the universe will reach infinity, so that any and all forces (no matter how strong) that hold composite objects together (no matter how closely) will be overcome by this expansion, literally tearing everything apart.

1.5.3 Big Crunch: $\geq 10^2$ billion years from now

See also: Big Crunch

If the energy density of dark energy were negative or the universe were closed, then it would be possible that the expansion of the universe would reverse and the universe would contract towards a hot, dense state. This is a required element of oscillatory universe scenarios, such as the cyclic model, although a Big Crunch does not necessarily imply an oscillatory universe. Current observations suggest that this model of the universe is unlikely to be correct, and the expansion will continue or even accelerate.

1.5.4 Big Freeze: $\geq 10^5$ billion years from now

Main articles: Future of an expanding universe and Heat death of the universe

This scenario is generally considered to be the most likely, as it occurs if the universe continues expanding as it has been. Over a time scale on the order of 10^{14} years or less, existing stars burn out, stars cease to be created, and the universe goes dark.[42], §IID. Over a much longer time scale in the eras following this, the galaxy evaporates as the stellar remnants comprising it escape into space, and black holes evaporate via Hawking radiation.[42], §III, §IVG. In some grand unified theories, proton decay after at least 10^{34} years will convert the remaining interstellar gas and stellar remnants into leptons (such as positrons and electrons) and photons. Some positrons and electrons will then recombine into photons.[42], §IV, §VF. In this case, the universe has reached a high-entropy state consisting of a bath of particles and low-energy radiation. It is not known however whether it eventually achieves thermodynamic equilibrium.[42], §VIB, VID.

1.5.5 Heat Death: 10^{1000} years from now

See also: Heat death of the universe

The heat death is a possible final state of the universe, estimated at after 10^{1000} years, in which it has "run down" to a state of no thermodynamic free energy to sustain motion or life. In physical terms, it has reached maximum entropy (because of this, the term "entropy" has often been confused with heat death, to the point of entropy being labelled as the "force killing the universe"). The hypothesis of a universal heat death stems from the 1850s ideas of William Thomson (Lord Kelvin)[43] who extrapolated the theory of heat views of mechanical energy loss in nature, as embodied in the first two laws of thermodynamics, to universal operation.

1.5.6 Vacuum metastability event

See also: False vacuum

If our universe is in a very long-lived false vacuum, it is possible that a small region of the universe will tunnel into a lower energy state (see Bubble nucleation). If this happens, all structures within will be destroyed instantaneously and the region will expand at near light speed, bringing destruction without any forewarning.

1.6   See also

- Cosmic Calendar (age of universe scaled to a single year)

- Cyclic model

- Dark-energy-dominated era

- Dyson's eternal intelligence

- Entropy (arrow of time)

- Graphical timeline from Big Bang to Heat Death

- Graphical timeline of the Big Bang

- Graphical timeline of the Stelliferous Era

- Illustris project

- Matter-dominated era

- Radiation-dominated era

- Timeline of the far future

- Ultimate fate of the universe

1.7   References

[1] Planck Collaboration (2015). "Planck 2015 results. XIII. Cosmological parameters (See Table 4 on page 31 of pfd).". arXiv:1502.01589. Bibcode:2015arXiv150201589P.

[2] Staff (17 March 2014). "BICEP2 2014 Results Release". *National Science Foundation*. Retrieved 18 March 2014.

[3] Clavin, Whitney (17 March 2014). "NASA Technology Views Birth of the Universe". *NASA*. Retrieved 17 March 2014.

[4] Overbye, Dennis (17 March 2014). "Detection of Waves in Space Buttresses Landmark Theory of Big Bang". *The New York Times*. Retrieved 17 March 2014.

[5] Chaisson, E., (2001). *Cosmic Evolution: The Rise of Complexity in Nature*, Harvard University Press, ISBN 0-674-00987-8; see also Cosmic Evolution

[6] Ryden B: "Introduction to Cosmology", pg. 196 Addison-Wesley 2003

[7] Overbye, Dennis (March 24, 2014). "Ripples From the Big Bang". *New York Times*. Retrieved March 24, 2014.

[8] Ade, P.A.R. (BICEP2 Collaboration); et al. (June 19, 2014). "Detection of B-Mode Polarization at Degree Angular Scales by BICEP2" (PDF). *Physical Review Letters* **112**: 241101. arXiv:1403.3985. Bibcode:2014PhRvL.112x1101A. doi:10.1103/Phys RevLett.112.241101.PMID24996078.Retrieved June20,2014.

[9] http://www.math.columbia.edu/~{}woit/wordpress/?p=6865

[10] Overbye, Dennis (June 19, 2014). "Astronomers Hedge on Big Bang Detection Claim". *New York Times*. Retrieved June 20, 2014.

[11] Amos, Jonathan (June 19, 2014). "Cosmic inflation: Confidence lowered for Big Bang signal". *BBC News*. Retrieved June 20, 2014.

[12] BICEP2/Keck, Planck Collaborations (2015). "A Joint Analysis of BICEP2/Keck Array and Planck Data (Provisionally accepted by PRL)". *arXiv*. arXiv:1502.00612v1. Retrieved 13 February 2015.

[13] Clavin, Whitney (30 January 2015). "Gravitational Waves from Early Universe Remain Elusive". *NASA*. Retrieved 30 January 2015.

[14] Overbye, Dennis (30 January 2015). "Speck of Interstellar Dust Obscures Glimpse of Big Bang". *New York Times*. Retrieved 31 January 2015.

[15] "Gravitational waves from early universe remain elusive". *Science Daily*. 31 January 2015. Retrieved 3 February 2015.

[16] The Timescale of Creation

[17] Detailed timeline of Big Bang nucleosynthesis processes

[18] Gannon, Megan (December 21, 2012). "New 'Baby Picture' of Universe Unveiled". Space.com. Retrieved December 21, 2012.

[19] Bennett, C.L.; Larson, L.; Weiland, J.L.; Jarosk, N.; Hinshaw, N.; Odegard, N.; Smith, K.M.; Hill, R.S.; Gold, B.; Halpern, M.; Komatsu, E.; Nolta, M.R.; Page, L.; Spergel, D.N.; Wollack, E.; Dunkley, J.; Kogut, A.; Limon, M.; Meyer, S.S.; Tucker, G.S.; Wright, E.L. (December 20, 2012). "Nine-Year Wilkinson Microwave Anisotropy Probe (WMAP) Observations: Final Maps and Results". *The Astrophysical Journal Supplement Series* **208**: 20. arXiv:1212.5225. Bibcode:2013ApJS..208...20B. doi:10.1088/0067-0049/208/2/20. Retrieved December 22, 2012.

[20] Hinshaw, G.; et al. (2009). "Five-Year Wilkinson Microwave Anisotropy Probe (WMAP) Observations: Data Processing, Sky Maps, and Basic Results" (PDF). *Astrophysical Journal Supplement* **180** (2): 225–245. arXiv:0803.0732. Bibcode:2009ApJS.. 180..225H.doi:10.1088/0067-0049/180/2/225.

[21] Mukhanov, V: "Physical foundations of Cosmology", pg. 120, Cambridge 2005

[22] Amos, Jonathan (2012-11-13). "Quasars illustrate dark energy's roller coaster ride". *BBC News*. Retrieved 13 November 2012.

[23] Loeb, Abraham (October 2014). "The Habitable Epoch of the Early Universe". *International Journal of Astrobiology* **13** (04): 337–339. arXiv:1312.0613. Bibcode:2014IJAsB..13..337L. doi:10.1017/S1473550414000196. Retrieved 15 December 2014.

[24] Loeb, Abraham (2 December 2013). "The Habitable Epoch of the Early Universe" (PDF). *Arxiv*. arXiv:1312.0613v3. Retrieved 15 December 2014.

[25] Dreifus, Claudia (2 December 2014). "Much-Discussed Views That Go Way Back - Avi Loeb Ponders the Early Universe, Nature and Life". *New York Times*. Retrieved 3 December 2014.

[26] Wall, Mike (December 12, 2012). "Ancient Galaxy May Be Most Distant Ever Seen". Space.com. Retrieved December 12, 2012.

[27] *Ferreting Out The First Stars*; physorg.com

[28]

[29] APOD: 2007 September 6 - Time Tunnel

[30] "New Scientist" 14 July 2007

[31] HET Helps Astronomers Learn Secrets of One of Universe's Most Distant Objects

[32] Scientists confirm most distant galaxy ever

[33] APOD: 2004 March 9 – The Hubble Ultra Deep Field

[34] Eduardo F. del Peloso a1a, Licio da Silva a1, Gustavo F. Porto de Mello and Lilia I. Arany-Prado (2005), "The age of the Galactic thin disk from Th/Eu nucleocosmochronology: extended sample" (Proceedings of the International Astronomical Union (2005), 1: 485-486 Cambridge University Press)

[35] "Cosmic Detectives". The European Space Agency (ESA). 2013-04-02. Retrieved 2013-04-15.

[36] K. P. Schroder, Robert Connon Smith (2008). "Distant future of the Sun and Earth revisited". *Monthly Notices of the Royal Astronomical Society* **386** (1): 155–163. arXiv:0801.4031. Bibcode:2008MNRAS.386..155S. doi:10.1111/j.1365-2966.2008.13022.x.

[37] J. Laskar (1994). "Large-scale chaos in the solar system". *Astronomy and Astrophysics* **287**: L9–L12. Bibcode:1994A&A...287L

[38] Zeilik & Gregory 1998, p. 320–321.

[39] "Introduction to Cataclysmic Variables (CVs)". *NASA Goddard Space Center*. 2006. Retrieved 2006-12-29.

[40] Palmer, Jason (22 February 2008). "Hope dims that Earth will survive Sun's death". *New Scientist*.

[41] G. Fontaine, P. Brassard, P. Bergeron (2001). "The Potential of White Dwarf Cosmochronology". *Publications of the Astronomical Society of the Pacific* **113** (782): 409–435. Bibcode:2001PASP..113..409F. doi:10.1086/319535. Retrieved 2008-05-11.

[42] A dying universe: the long-term fate and evolution of astrophysical objects, Fred C. Adams and Gregory Laughlin, *Reviews of Modern Physics* **69**, #2 (April 1997), pp. 337–372. Bibcode: 1997RvMP...69..337A. doi:10.1103/RevModPhys.69.337.

[43] Thomson, William. (1851). "On the Dynamical Theory of Heat, with numerical results deduced from Mr Joule's equivalent of a Thermal Unit, and M. Regnault's Observations on Steam." Excerpts. [§§1-14 & §§99-100], *Transactions of the Royal Society of Edinburgh*, March, 1851; and *Philosophical Magazine* IV. 1852, [from *Mathematical and Physical Papers*, vol. i, art. XLVIII, pp. 174]

1.8 External links

- PBS Online (2000). From the Big Bang to the End of the Universe – The Mysteries of Deep Space Timeline. Retrieved March 24, 2005.

- Schulman, Eric (1997). The History of the Universe in 200 Words or Less. Retrieved March 24, 2005.

- Space Telescope Science Institute Office of Public Outreach (2005). Home of the Hubble Space Telescope. Retrieved March 24, 2005.

- Fermilab graphics (see "Energy time line from the Big Bang to the present" and "History of the Universe Poster")

- Exploring Time from Planck time to the lifespan of the Universe

- Cosmic Evolution is a multi-media web site that explores the cosmic-evolutionary scenario from big bang to humankind.

- Astronomers' first detailed hint of what was going on less than a trillionth of a second after time began

- The Universe Adventure

- Cosmology FAQ, Professor Edward L. Wright, UCLA

- Sean Carroll on the arrow of time (Part 1), *The origin of the universe and the arrow of time*, Sean Carroll, video, CHAST 2009, Templeton, Faculty of science, University of Sydney, November 2009, TED.com

- A Universe From Nothing, video, Lawrence Krauss, AAI 2009, YouTube.com

- Once Upon A Universe - Story of the Universe told in 13 chapters. Science communication site supported by STFC.

- Cosmic Evolution through Time - an interactive timeline explains the main events in the history of our Universe

Chapter 2

Timeline of the formation of the Universe

Main article: Chronology of the universe
This is a timeline of the formation and subsequent evolution of the Universe from the Big Bang 13.799 ± 0.021 billion

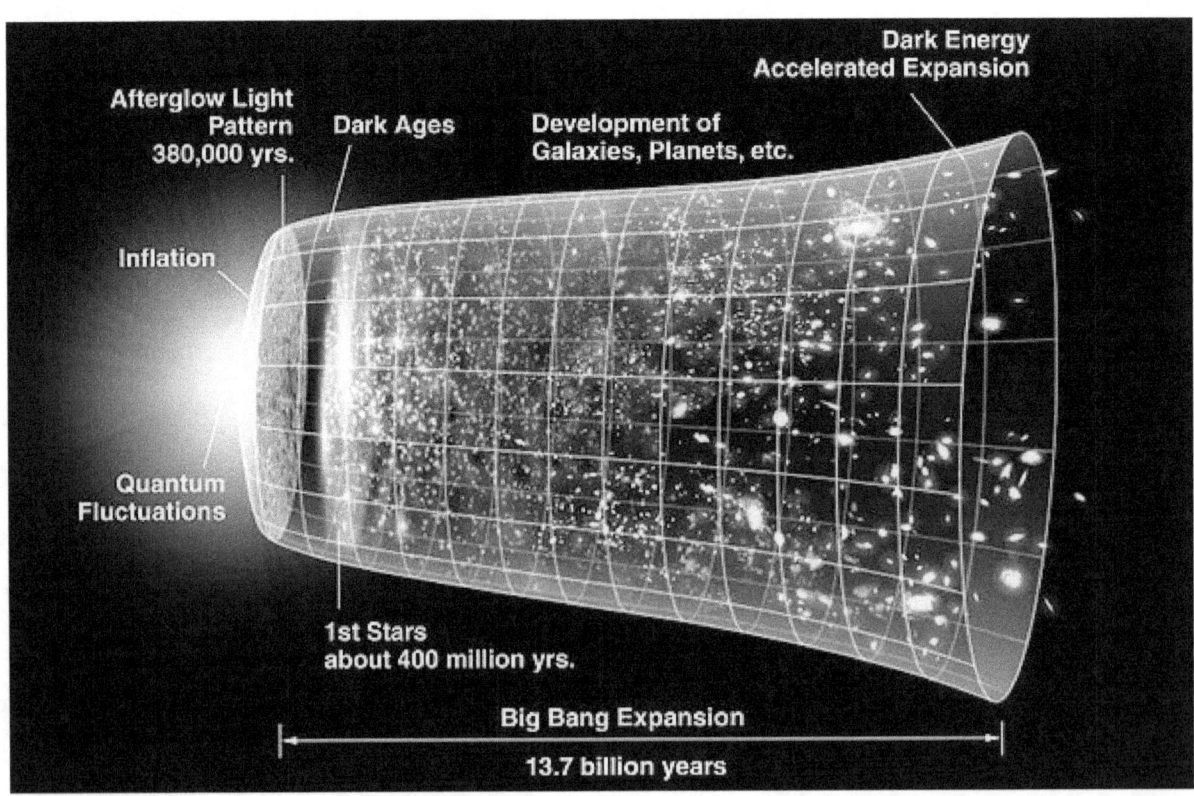

Diagram of Evolution of the universe from the Big Bang (left) - to the present.

years ago to the present day. Times are measured from the moment of the Big Bang.

2.1 The first second

2.1.1 Planck Epoch

- ca. 0 seconds (13.799 ± 0.021 Gya): Planck Epoch begins: earliest meaningful time. The Big Bang occurs in which ordinary space and time develop out of a primeval state (possibly a virtual particle or false vacuum) described by a quantum theory of gravity or "Theory of Everything". All matter and energy of the entire visible Universe is contained in an unimaginably hot, dense point (Gravitational singularity), a billionth the size of a nuclear particle. This state has been described as a particle desert. Other than a few scant details conjecture dominates discussion about the earliest moments of the universe's history since no effective means of testing this far back in space-time is presently available. WIMPS (weakly interacting massive particles) or dark matter and dark energy may have appeared and been the catalyst for the expansion of the singularity. Infant Universe experiences cooling as it begins expanding outward - almost completely smooth, quantum variations begin causing slight variations in density

Grand Unification Epoch

- ca. 10^{-43} seconds: Grand unification epoch begins: While still at an infinitesimal size, Universe cools down to 10^{32} Kelvin. Gravity separates and begins operating on the Universe - remaining fundamental forces stabilize into electronuclear force, also known as the Grand Unified Force or Grand Unified Theory (GUT). Hypothetical X and Y bosons[1] appear however physical characteristics such as mass, charge, flavour and colour charge are meaningless.

2.1.2 Electroweak epoch

- ca. 10^{-36} seconds: Electroweak epoch begins: The Universe cools down to 10^{28} Kelvin. As a result, the Strong Nuclear Force becomes distinct from the Electroweak Force perhaps fuelling the inflation (cosmology) of the universe. A wide array of exotic elementary particles result from decay of X and Y bosons which include W and Z bosons and Higgs bosons.

- ca. 10^{-33} seconds: Space is subjected to a superfast inflation, influenced by a replusive energy field, expanding from the size of an atom to that of a grapefruit in a tiny fraction of a second. The inflation also generates two types of waves (gravitational and density) along which the previous quantum fluctuations inflate becoming structures that will influence future galaxy clustering. Some postulations suggest that this inflation may be eternal and may have been responsible for a multiverse which contains our own universe. At present concept has been subject to serious debate throughout present-day scientific community and is a matter of conjecture

- ca. 10^{-32} seconds: Cosmic inflation ends. Particles of matter (quarks, gluons, electrons) form as a soup of hot ionized gas called quark-gluon plasma - photons of light (radiation are scattered by travelling through this plasma. It is also possible that one potential type of dark matter (axions) is synthesized.

2.1.3 Quarks Epoch

- ca. 10^{-12} seconds: Quark epoch begins. Electroweak phase transition: supersymmetry breaking as the Electromagnetic and Weak Nuclear forces become distinct. Cooling weakens the Weak nuclear force, so matter particles can acquire mass and interact with the Higgs Field. As fundamental interactions begin acting on the Universe, it remains too hot for quarks to bind together into larger forms of matter - domination of radiation over matter with quarks and gluons experiencing degrees of freedom. The universe cools to 10^{15} Kelvin.

- ca. 10^{-11} seconds: Baryogenesis may have taken place with matter gaining the upper hand over anti-matter as baryon to antibaryon constituencies are established. A second potential type of dark matter (neutrinos) may have been synthesized.

2.1.4 Hadron Epoch

- ca. 10^{-6} seconds: Hadron epoch begins: As the universe cools to about 10^{10} Kelvin, a quark-hadron transition takes place in which quarks bind to form more complex particles - hadrons. This quark confinement includes the formation of protons and neutrons (nucleons), the building blocks of atomic nuclei.

- ca. 1 second: Lepton epoch begins: The universe cools to 10^9 Kelvin. At this temperature, the hadrons and antihadrons annihilate each other, leaving behind leptons and antileptons - possible disappearance of antiquarks. Gravity governs the expansion of the universe: neutrinos decouple from matter creating a cosmic neutrino background.

2.2 Matter Era

2.2.1 Photon Epoch

- ca. 10 seconds: Photon epoch begins: Most of the leptons and antileptons annihilate each other. As electrons and positrons annihilate, a small number of unmatched electrons are left over - disappearance of the positrons.

- ca. 10 seconds: Universe dominated by photons of radiation - ordinary matter particles are coupled to light and radiation while dark matter particles start building non-linear structures as dark matter halos. Because charged electrons and protons hinder the emission of light, the universe becomes a super-hot glowing fog.

- ca. 3 minutes: Primordial Nucleosynthesis: nuclear fusion begins as lithium and heavy hydrogen (deuterium) and helium nuclei form from protons and neutrons.

- ca. 20 minutes: Nuclear fusion ceases: normal matter consists of 75% hydrogen and 25% helium - free electrons begin scattering light.

- ca. 70,000 yrs: Matter domination in Universe: onset of gravitational collapse as the Jeans Length at which the smallest structure can form begins to fall.

2.2.2 Cosmic Dark Age

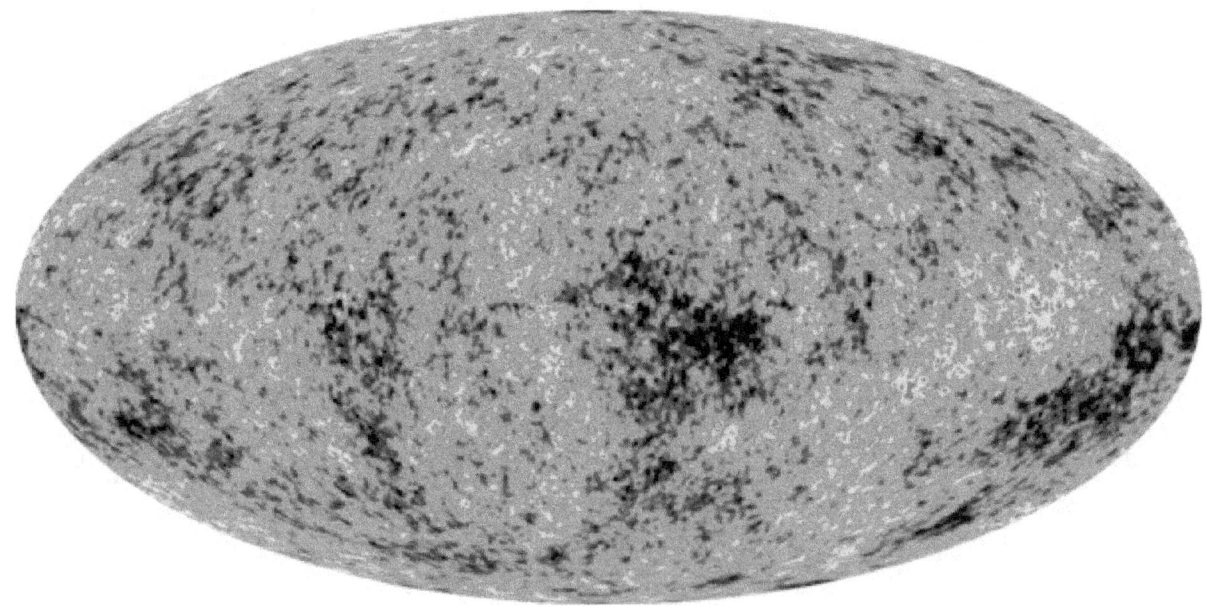

All-sky map of the CMB, created from 9 years of WMAP data.

Main article: List of cosmic microwave background experiments

- ca. 370,000 yrs: Dark Ages (cosmology) begin - Recombination: electrons combine with nuclei to form atoms mostly hydrogen and helium. The glow from our infant Universe is unveiled. Distributions of hydrogen and helium at this time remains constant as the electron-baryon plasma thins. The temperature falls to 3000 degrees Kelvin. Ordinary matter particles decouple from radiation. The photons fly free releasing a Cosmic Microwave Background. The universe becomes neutral and transparent. the Afterglow light pattern source taken by later satellites is the farthest back our instruments can see. The Cosmic Microwave Background is also the only light source: with no stars there is no other source of light. The Universe empty except for the neutral clouds of hydrogen and helium.

- ca. 400,000 yrs: Density waves begin imprinting characteristic polarization (waves) signals.

- ca. 10 million yrs: With a trace of heavy elements in the Universe, Abiogenesis (chemistry of life) begins operating.

- ca. 100 million yrs: Gravitational collapse: ordinary matter particles fall into the structures created by dark matter. Reionization begins: smaller (stars) and larger non-linear structures (quasars) begin to take shape - their ultraviolet light ionizes remaining neutral gas

- 200-300 million yrs: First stars begin to shine: Because many are Population III stars (some Population II stars are accounted for at this time) they are much bigger and hotter and their life-cycle is fairly short. Unlike later generations of stars, these stars are metal free. As reionization intensifies, photons of light scatter off free protons and electrons - Universe becomes opaque again

- 200 million years: HD 140283, the "Methuselah" Star, formed, the unconfirmed oldest star observed in the Universe. Because it is a Population II star, some suggestions have been raised that second generation star formation may have begun very early on.[2] The oldest known star (confirmed) - SMSS J031300.36-670839.3, forms.

- 300 million years First large-scale astronomical objects, protogalaxies and quasars may have begun forming. As Population III stars continue to burn, stellar nucleosynthesis operates - stars burn mainly by fusing hydrogen to produce more helium in what is referred to as the Main Sequence. Over time these stars are forced to fuse helium to produce carbon, oxygen, silicon and other heavy elements up to iron on the periodic table. These elements, when seeded into neighbouring gas clouds by supernova, will lead to the formation of more Population II stars (metal poor) and gas giants.

- 380 million yrs: UDFj-39546284, current record holder for oldest known quasar.[3]

- 420 million yrs: The quasar MACS0647-JD, forms

- 470 - 500 million yrs: Abell 1835 IR1916 forms.

2.2.3   Renaissance

- 600 million yrs: Renaissance of the Universe - end of the Dark Ages as visible light begins dominating throughout. Possible formation of the Milky Way Galaxy: although age of Methusaleh star suggests a much older date of origin, it is highly likely that HD 140283 may have come into our galaxy via a later galaxy merger. Oldest confirmed star in Milky Way Galaxy, HE 1523-0901. Extent of the Hubble Extreme Deep Field.

- 630 million yrs: GRB 090423, the oldest gamma ray burst recorded suggests that supernovas may have happened very early on in the evolution of the Universe[4]

- 670 million yrs: EGS-zs8-1, the most distant starburst or Lyman-break galaxy observed, forms. This suggests that galaxy interaction is taking place very early on in the history of the Universe as starburst galaxies are often associated with collisions and galaxy mergers.

- 700 million yrs: Galaxies form. Smaller galaxies begin merging to form larger ones. Galaxy classes may have also begun forming at this time including Blazars, Seyfert galaxies, radio galaxies, normal galaxies (elliptical, Spiral galaxies, barred spiral) and dwarf galaxies. UDFy-38135539, the first distant quasar to be observed from the reionization phase, forms. Dwarf galaxy z8 GND 5296 forms. Galaxy or possible proto-galaxy A1689-zD1 forms.

- 720 million yrs: Possible formation of globular clusters in Milky Way's Galactic halo. Formation of globular cluster, NGC 6723, in the Milky Way's galactic halo

- 740 million years: 47 Tucanae, second brightest globular cluster in the Milky Way, forms

- 750 million yrs: Galaxy IOK-1 a Lyman alpha emitter galaxy, forms. GN-108036 forms - galaxy is 5 times larger and 100 times more massive than the present day Milky Way illustrating the size attained by some galaxies very early on.

- 770 million yrs: Quasar ULAS J1120+0641, one of the most distant, forms. One of the earliest galaxies to feature a supermassive black hole suggesting that such large objects existed quite soon after the Big Bang. The large fraction of neutral hydrogen in its spectrum suggests it may also have just formed or is in the process of star formation.

- 800 million yrs: Farthest extent of Hubble Ultra Deep Field. Formation of SDSS J102915+172927: unusual population II star that is extremely metal poor consisting of mainly hydrogen and helium. HE0107-5240, one of the oldest Population II stars, forms as part of a binary star system. LAE J095950.99+021219.1, the Bogwiggit Galaxy, one of the most remote Lyman alpha emitter galaxies, forms. Lyman alpha emitters are considered to be the progenitors of spiral galaxies like the Milky Way. Messier 2, globular cluster, forms.

- 870 million yrs: Messier 30 forms in the Milky Way. Having experienced a Core collapse (cluster), the cluster has one of the highest densities among globular clusters.

- 890 million yrs: Galaxy SXDF-NB1006-2 forms

- 900 million yrs: Galaxy BDF-3299 forms.

- 910 million yrs: Galaxy BDF-521 forms

2.2.4 Galaxy Epoch

- 1 billion yrs (12.8 Gya): Galaxy HCM-6A, the most distant normal galaxy observed, forms. Formation of hyper-luminous quasar SDSS J0100+2802 - It harbors a black hole with mass of 12 billion solar masses one of the most massive black hole discovered so early in the universe. HE1327-2326, population II star, speculated to have formed from remnants of earlier Population III stars. Visual limit of the Hubble Deep Field. Reionization complete - the Universe becomes transparent again. Galaxy evolution continues as more modern looking galaxies form and develop. Because the Universe is still small in size, galaxy interactions become common place with larger and larger galaxies forming out of the galaxy merger process. Galaxies may have begun clustering creating the largest structures in the Universe so far - the first galaxy clusters and galaxy superclusters appear.

- 1.1 billion yrs (12.7 Gya): Age of the quasar CFHQS 1641+3755. Messier 4 Globular Cluster, first to have its individual stars resolved, forms in the halo of the Milky Way Galaxy. Among the clusters many stars, PSR B1620-26 b, a gas giant known as the "Genesis Planet" or "Methusaleh", orbiting a pulsar and a white dwarf, the oldest observed extrasolar planet in Universe, forms.

- 1.13 billion yrs (12.67 Gya): Messier 12, globular cluster, forms

- 1.3 billion yrs (12.5 Gya: WISE J224607.57-052635.0, a luminous infrared galaxy, forms. PSR J1719-1438 b, known as the Diamond Planet forms around a pulsar.

- 1.31 billion yrs (12.49 Gya): Globular Cluster Messier 53 forms 60,000 light-years from the galactic centre of the Milky Way

- 1.39 billion yrs (12.41 Gya): S5 0014+81, a hyper-luminous quasar, forms

- 1.4 billion yrs (12.4 Gya): Age of Cayrel's Star, BPS C531082-0001, a neutron capture star, among the oldest Population II stars in Milky Way. Quasar RD1, first object observed to exceed redshift 5, forms.

- 1.44 billion yrs (12.36 Gya): Messier 80 globular cluster forms in Milky Way - known for large number of "blue stragglers"

- 1.5 billion yrs (12.3 Gya): Messier 55, globular cluster, forms

- 1.8 billion yrs (12 Gya): Most energetic gamma ray burst lasting 23 minutes, GRB 080916C, recorded. Baby Boom Galaxy forms. Terzan 5 forms as a small dwarf galaxy on collision course with the Milky Way. Dwarf galaxy carrying the Methusaleh Star consumed by Milky Way - oldest known star in the Universe becomes one of many population II stars of the Milky Way

- 2.0 billion yrs (11.8 Gya): SN 1000+0216, the oldest observed supernova occurs - possible pulsar formed. Globular Cluster Messier 15, known to have an intermediate black hole and the only globular cluster observed to include a planetary nebula, Pease 1, forms

- 2.02 billion yrs (11.78 Gya): Messier 62 forms - contains high number of variable stars (89) many of which are RR Lyrae stars.

- 2.2 billion yrs (11.6 Gya): Globular Cluster NGC 6752, third brightest, forms in Milky Way

- 2.4 billion yrs (11.4 Gya): Quasar PKS 2000-330 forms.

- 2.41 billion yrs (11.39 Gya): Messier 10 globular cluster forms. Messier 3 forms: prototype for the Oosterhoff type I cluster, which is considered "metal-rich". That is, for a globular cluster, Messier 3 has a relatively high abundance of heavier elements.

- 2.5 billion yrs (11.3 Gya): Omega Centauri, largest globular cluster in the Milky Way forms

- 3.0 billion yrs (10.8 billion Gya): Formation of Gliese 581 planetary system: Gliese 581 c, the first observed ocean planet and Gliese 581 d, a super-earth planet, possibly the first observed habitable planets, form. Gliese 581 d has more potential for forming life since it is the first exoplanet of terrestrial mass proposed that orbits within the habitable zone of its parent star.

- 3.3 billion yrs (10.5 Gya): BX442, oldest grand design spiral galaxy observed, forms

- 3.5 billion yrs (10.3 Gya): Supernova SN UDS10Wil recorded

- 3.8 billion yrs (10 Gya): NGC 2808 globular cluster forms: 3 generations of stars form within the first 200 million years. Mu Cephei, giant red star forms

- 4.0 billion yrs (9.8 Gya): Quasar 3C 9 forms. The Andromeda galaxy forms from a galactic merger - begins a collision course with the Milky Way. Barnard's Star, red dwarf star, may have formed. Beethoven Burst GRB 991216 recorded. Gliese 677 Cc, a planet in the habitable zone of its parent star, Gliese 667, one of the most physically similar known exoplanets to Earth, forms

- 4.1 billion yrs (9.7 Gya): 16 Cygni Bb, the first gas giant observed in a single star orbit in a trinary star system, forms - orbiting moons considered to have habitable properties or at the least capable of supporting water

- 4.5 billion yrs (9.3 Gya): Fierce star formation in Andromeda making it into a luminous infra-red galaxy

- 5.0 billion yrs (8.8 Gya): Earliest Population I, or Sunlike stars: with heavy element saturation so high, planetary nebula appear in which rocky substances are solidified - these nurseries lead to the formation of rocky terrestrial planets, moons, asteroids, and icy comets

- 5.1 billion yrs (8.7 Gya): Galaxy collision: spiral arms of the Milky Way form leading to major period of star formation.

- 5.3 billion yrs (8.5 Gya): 55 Cancri B, a "hot Jupiter", first planet to be observed orbiting as part of a star system, forms. Kepler 11 planetary system, the flattest and most compact system yet discovered, forms - Kepler 11 c considered to be a giant ocean planet with hydrogen-helium atmosphere.

- 5.8 billion yrs (8 Gya): 51 Pegasi b also known as Bellerophon, forms - first planet discovered orbiting a main sequence star

- 5.9 billion yrs (7.9 Gya): HD 176051 planetary system, known as the first observed through astrometrics, forms

2.2.5 Acceleration

- 6.0 billion yrs (7.8 Gya): Acceleration: dark energy begins dominating Universe - after being slowed for billions of years by gravity abundant dark matter takes hold and the cosmic expansion begins to speed up. As the cosmic expansion accelerates, the rate of galaxy interactions decreases - although near misses continue to distort some while collisions increase the size of others, distance makes the galaxy merger process less likely. Many galaxies like NGC 4565 become relatively stable - ellipticals result from collisions of spirals with some like IC 1101 being extremely massive. Rigel or Beta Orionis, an alpha cygni variable, forms.

- 6.0 billion years-present (7.8 Gya-present): The Universe continues to organize into larger wider structures. The great walls, sheets and filaments consisting of galaxy clusters and superclusters and voids crystallize. How this crystallization takes place is still conjecture. Certainly, it is possible the formation of super-structures like the Hercules-Corona Borealis Great Wall may have happened much earlier, perhaps around the same time galaxies first started appearing. Either way the observable universe becomes more modern looking.

- 6.3 billion yrs (7.5 Gya): GRB 080319B, farthest gamma ray burst seen with the naked eye, recorded. Terzan 7, metal-rich globular cluster, forms in the Sagittarius Dwarf Elliptical Galaxy

- 6.5 billion yrs (7.3 Gya): HD 10180 planetary system forms (larger than both 55 Cancri and Kepler 11 systems)

- 6.9 billion yrs (6.9 Gya): Orange Giant, Arcturus, forms

- 7 billion yrs (6.8 Gya): North Star, Polaris, one of the significant navigable stars, forms

- 7.64 billion yrs (6.16 Gya): Mu Arae planetary system forms: of four planets orbiting a yellow star, Mu Arae c is among the first terrestrial planets to be observed from Earth

- 7.8 billion yrs (6 Gya): Formation of Earth's near twin, Kepler 452b orbiting its parent star Kepler 452

- 7.98 billion yrs (5.82 Gya): Formation of Mira or Omicron ceti, binary star system. Formation of Alpha Centauri Star System, closest star to the Sun - formation of Alpha Centauri Bb closest planet to the Sun. GJ 1214 b, or Gliese 1214 b, potential earth-like planet, forms

- 8.08-8.58 billion yrs (5.718-5.218 Gya): Capella star system forms

- 8.2 billion yrs (5.6 Gya): Tau Ceti, nearby yellow star forms: five planets eventually evolve from its planetary nebula, orbiting the star - Tau Ceti e considered planet to have potential life since it orbits the hot inner edge of the star's habitable zone

- 8.5 billion yrs (5.3 Gya): GRB 101225A, the "Christmas Burst", considered the longest at 28 minutes, recorded

- 8.8 billion yrs (5 Gya): Messier 67 open star cluster forms: Three exoplanets confirmed orbiting stars in the cluster including a twin of our Sun

- 9.0 billion yrs (4.8 Gya): Lalande 21185, red dwarf in Ursa Major, forms

- 9.13 billion yrs (4.67 Gya): Proxima Centauri forms completing the Alpha Centauri binary system

2.3 Solar Era

- 9.2 billion yrs (4.6 Gya): Primal supernova, possibly triggers the formation of the Solar System.

- 9.231 billion yrs (4.568 Gya): Sun forms - Planetary nebula begins accretion of planets.

- 9.25 billion yrs (4.55 Gya): Solar System of Eight planets, four terrestrial (Mercury (planet), Venus, Earth, Mars) and four Jovian planets (Jupiter. Saturn, Uranus, Neptune) evolve around the Sun. Because of accretion many smaller planets form orbits around the proto-Sun some with conflicting orbits - Early Bombardment Phase begins. Pre-Noachian Era begins on Mars. Pre-Tolstojan Period begins on Mercury - Large planetoid strikes Mercury stripping it of outer envelope of original crust and mantle, leaving the planet's core exposed - Mercury's iron content notably high. Vega, fifth brightest star in our galactic neighbourhood, forms. Many of the Galilean moons may have formed at this time including Europa and Titan which may presently be hospitable to some form of living organism.

- 9.254 billion yrs (4.545 Gya): Major collision with a planetoid establishes the Martian dichotomy on Mars - formation of North Polar Basin (Mars)

- 9.266 billion yrs (4.533 Gya): Formation of Earth-Moon system following giant impact by hypothetical planetoid Thea (planet). Moon's gravitational pull helps stabilize Earth's fluctuating axis of rotation. Pre-Nectarian Period begins on Moon

- 9.3 billion yrs (4.5 Gya): Sun becomes a main sequence yellow star: formation of the Oort Cloud and Kuiper Belt from which a stream of comets like Halley's Comet and Hale-Bopp begins passing through the Solar System, sometimes colliding with planets and the Sun

- 9.4 billion yrs (4.4 Gya): Formation of Kepler 438 b, one of the most Earth-like planets, from a protoplanetary nebula surrounding its parent star

- 9.5 billion yrs (4.3 Gya): Massive meteorite impact creates South Pole Aitken Basin on the Moon - a huge chain of mountains located on the lunar southern limb, sometimes called "Leibnitz mountains", form

- 9.6 billion yrs (4.2 Gya): Tharsis Bulge widespread area of vulcanism, becomes active on Mars - based on the intensity of volcanic activity on Earth, Tharsis magmas may have produced a 1.5-bar CO_2 atmosphere and a global layer of water 120 m deep increasing greenhouse gas effect in climate and adding to Martian water table. Age of the oldest samples from the Lunar Maria

- 9.7 billion yrs (4.1 Gya): Resonance in Jupiter and Saturn's orbits moves Neptune out into the Kuiper belt causing a disruption among asteroids and comets there. As a result, Late Heavy Bombardment batters the inner Solar System. Herschel Crater formed on Mimas (moon), a moon of Saturn. Meteorite impact creates the Hellas Planitia on Mars, the largest unambiguous structure on the planet. Anseris Mons an isolated massif (mountain) in the southern highlands of Mars, located at the northeastern edge of Hellas Planitia is uplifted in the wake of the meteorite impact

- 9.8 billion yrs (4 Gya): HD 209458 b, first planet detected through its transit, forms. Messier 85, lenticular galaxy, disrupted by galaxy interaction: complex outer structure of shells and ripples results. Andromeda and Triangulum galaxies experience close encounter - high levels of star formation in Andromeda while Triangulum's outer disc is distorted

- 9.861 billion yrs (3.938 Gya): Major period of impacts on the Moon: Mare Imbrium forms

- 9.88 billion yrs (3.92 Gya): Nectaris Basin forms from large impact event: ejecta from Nectaris forms upper part of densely cratered Lunar Highlands - Nectarian Era begins on the Moon.

- 9.9 billion yrs (3.9 Gya): Tolstoj (crater) forms on Mercury. Caloris Basin forms on Mercury leading to creation of "Weird Terraine" - seismic activity triggers volcanic activity globally on Mercury. Rembrandt (crater) formed on Mercury. Caloris Period begins on Mercury. Argyre Planitia forms from asteroid impac t on Mars: surrounded by rugged massifs which form concentric and radial patterns around basin - several mountain ranges including Charitum and Nereidum Montes are uplifted in its wake

- 9.95 billion yrs (3.85 Gya): Beginning of Late Imbrium Period on Moon. Earliest appearance of Procellarum KREEP Mg suite materials

- 9.96 billion yrs (3.84 Gya): Formation of Orientale Basin from asteroid impact on Lunar surface - collision causes ripples in crust, resulting in three concentric circular features known as Montes Rook and Montes Cordillera

2.4   Biological Era

- 10 billion years (3.8 Gya): In the wake of Late Heavy Bombardment impacts on the Moon, large molten mare depressions dominate lunar surface - major period of Lunar vulcanism begins (to 3 Gyr)

- 10.2 billion years (3.6 Gya): Alba Mons forms on Mars, largest volcano in terms of area

- 10.6 billion years (3.2 Gya): Amazonian (Mars) Period begins on Mars: Martian climate thins to its present density: groundwater stored in upper crust (megaregolith) begins to freeze, forming thick cryosphere overlying deeper zone of liquid water - dry ices composed of frozen carbon dioxide form Eratosthenian period begins on the Moon: main geologic force on the Moon becomes impact cratering

- 10.8 billion years (3 Gya): Beethoven Basin forms on Mercury - unlike many basins of similar size on the Moon, Beethoven is not multi ringed and ejecta buries crater rim and is barely visible

- 11.6 billion yrs (2.2 Gya): Last great tectonic period in Martian geologic history: Valles Marineris, largest canyon complex in the Solar System, forms - although some suggestions of thermokarst activity or even water erosion, it is suggested Valles Marineris is rift fault

- 11.8 billion yrs (2 Gya): Star formation in Andromdea Galaxy slows. Formation of Hoag's Object from a galaxy collision. Olympus Mons largest volcano in the Solar System forms

- 12.1 billion yrs (1.7 Gya): Sagittarius Dwarf Elliptical Galaxy captured into an orbit around Milky Way Galaxy

- 12.7 billion yrs (1.1 Gya): Copernican Period begins on Moon: defined by impact craters that possess bright optically immature ray systems

- 12.8 billion yrs (1 Gya): Kuiperian Era (1 Gyr -) begins on Mercury: modern Mercury, desolate cold planet influenced by space erosion and solar wind extremes. Interactions between Andromeda and its companion galaxies Messier 32 and Messier 110. Galaxy collision with Messier 82 forms its spiral patterned disc: galaxy interactions between NGC 3077 and Messier 81

- 13 billion yrs (800 Mya): Copernicus (lunar crater) forms from impact on Lunar surface in the area of Oceanus Procellarum - has terrace inner wall and 30 km wide, sloping rampart that descends nearly a kilometer to the surrounding mare

- 13.0-13.4 billion yrs (0.8-0.4 Gya): Epsilon Eridani, third closest star to the Sun forms - From its planetary nebula Epsilon eridani b (gas giant) forms

- 13.175 billion yrs (625 Mya): formation of Hyades star cluster: consists of a roughly spherical group of hundreds of stars sharing same age, place of origin, chemical content and motion through space

- 13.2 billion yrs (600 Mya): Collision of spiral galaxies leads to creation of Antenna Galaxies. Whirlpool Galaxy collides with NGC 5195 forming present connected galaxy system. HD 189733 b forms around parent star HD 189733: first planet to reveal climate, organic constituencies, even colour (blue) of its atmosphere

- 13.5-13.6 billion yrs (200-300 Mya): Sirius, the brightest star in the Earth's sky, forms.

- 13.787 billion yrs (12 Mya): Antares forms.

- 13.791 billion yrs (7.6 Mya): Betelgeuse forms.

- 13.795 billion yrs (4.4 Mya): Fomalhaut b, first directly imaged planet, forms

- 13.799 billion yrs: Present day.

2.5   See also

- Timelines of world history

- Timeline of the far future

2.6 References

[1] Ta-Pei Cheng; Ling-Fong Li (1983). *Gauge Theory of Elementary Particle Physics*. Oxford University Press. p. 437. ISBN 0-19-851961-3.

[2] R. Cowen (January 10, 2013). "Nearby star is almost as old as the Universe". Nature News. doi:10.1038/nature.2013.12196. Retrieved February 23, 2013.

[3] Wall, Mike (December 12, 2012). "Ancient Galaxy May Be Most Distant Ever Seen". Space.com. Retrieved December 12, 2012.

[4] "GRB 090423 goes Supernova in a galaxy, far, far away". *Zimbio*. Retrieved 2010-02-23.

Chapter 3

Timeline of fundamental physics discoveries

250 BCE	Archimedes' principle: Archimedes
1514	Heliocentrism: Nicholas Copernicus
1589	Galileo's Leaning Tower of Pisa experiment: Galileo Galilei
1600	Earth's magnetic field discovered : William Gilbert
1613	Inertia: Galileo Galilei
1621	Snell's law: Willebrord Snellius
1660	Pascal's Principle: Blaise Pascal
1660	Hooke's law: Robert Hooke
1687	Laws of motion and law of gravity: Newton
1780	
1782	Conservation of matter: Lavoisier
1785	Inverse square law for electric charges confirmed: Coulomb
1801	Wave theory of light: Young
1803	Atomic theory of matter: Dalton
1805	
1806	Kinetic energy: Young
1814	Wave theory of light, interference: Fresnel
1820	Evidence for electromagnetic interactions: Ampère, Biot, Savart
1824	Ideal gas cycle analysis, internal combustion engine: Sadi Carnot
1827	Electrical resistance, etc.: Ohm
1830	
1838	Lines of force, fields: Michael Faraday
1838	Earth's magnetic field: Weber and Gauss
1842–3	Conservation of energy: Mayer, Kelvin
1842	Doppler effect: Kelvin
1845	Faraday rotation (light and electromagnetic): Faraday
1847	Conservation of energy 2: Joule, Helmholtz
1850–1	Second law of thermodynamics: Clausius, Kelvin
1855	
1857–9	Kinetic theory: Clausius, Maxwell
1861	Black body: Kirchhoff
1863	Entropy: Clausius
1864	Dynamical theory of the electromagnetic field: Maxwell
1867	Dynamic theory of gases: Maxwell
1880	
1871–89	Statistical mechanics: Boltzmann, Gibbs
1884	Boltzmann derives Stefan's radiation law
1887	Electromagnetic waves: Hertz
1893	Radiation law: Wien
1895	X-rays: Röntgen
1896	Radioactivity: Becquerel
1897	Electron: Thomson
1900	Formula for Black body radiation: Planck
1905	
1905	Special relativity: Einstein Photoelectric effect: Einstein Brownian motion: Einstein
1911	Equivalence principle Discovery of the atomic nucleus: Rutherford Superconductivity: Kamerlingh Onnes
1913	Bohr model of the atom: Bohr
1916	General relativity: Einstein
1922	Friedmann proposes expanding universe
1923	Stern–Gerlach experiment Matter waves Galaxies Particle nature of photons confirmed
1925	Stellar structure understood
1927	Big Bang: Lemaître
1928	Antimatter predicted: Dirac
1929	Expansion of universe confirmed: Hubble
1932	Antimatter discovered: Anderson Neutron discovered: Chadwick
1937	Muon discovered: Anderson & Neddermeyer
1938	Superfluidity discovered Nuclear fission discovered
1947	Pion discovered
1948	Theory of Quantum electrodynamics
1956	Electron neutrino discovered

1956–7	Parity violation discovered
1957	Theory of Superconductivity
1962	Theory of strong interactions Muon neutrino discovered
1967	Theory of Weak Interaction Pulsars discovered
1974	Charmed quark discovered
1975	Tau lepton discovered
1977	Bottom quark discovered
1980	Quantum Hall effect discovered
1981	Theory of cosmic inflation Fractional quantum Hall effect discovered
1995	Top quark discovered
1998	Accelerating universe discovered
2000	Tau neutrino discovered
2012	Higgs Boson discovered
2016	Gravitational waves detected

Chapter 4
Timeline of classical mechanics

Timeline of classical mechanics:

4.1 Early Mechanics

- 4th century BC - Aristotle founds the system of Aristotelian physics

- 260 BC - Archimedes mathematically works out the principle of the lever and discovers the principle of buoyancy

- 60 AD - Hero of Alexandria writes *Metrica, Mechanics,* and *Pneumatics*

- 1021 - Al-Biruni realizes that acceleration is connected with non-uniform motion[1]

- 1000-1030 - Alhazen and Avicenna develop the concepts of inertia and momentum

- 1100-1138 - Avempace develops the concept of a reaction force[2]

- 1100-1165 - Hibat Allah Abu'l-Barakat al-Baghdaadi discovers that force is proportional to acceleration rather than speed, a fundamental law in classical mechanics[3]

- 1121 - Al-Khazini publishes *The Book of the Balance of Wisdom*, in which he develops the concepts of gravitational potential energy and gravity at-a-distance[4]

- 1340-1358 - Jean Buridan develops the theory of impetus

- 1490 - Leonardo da Vinci describes capillary action

- 1500-1528 - Al-Birjandi develops the theory of "circular inertia" to explain Earth's rotation[5]

- 1581 - Galileo Galilei notices the timekeeping property of the pendulum

- 1589 - Galileo Galilei uses balls rolling on inclined planes to show that different weights fall with the same acceleration

- 1638 - Galileo Galilei publishes *Dialogues Concerning Two New Sciences*

- 1658 - Christiaan Huygens experimentally discovers that balls placed anywhere inside an inverted cycloid reach the lowest point of the cycloid in the same time and thereby experimentally shows that the cycloid is the tautochrone

- 1668 - John Wallis suggests the law of conservation of momentum

- 1676-1689 - Gottfried Leibniz develops the concept of *vis viva*, a limited theory of conservation of energy

4.2 Formation of Classical Mechanics (sometimes referred to as Newtonian mechanics)

- 1687 - Isaac Newton publishes his *Philosophiae Naturalis Principia Mathematica*, in which he formulates Newton's laws of motion and Newton's law of universal gravitation

- 1690 - James Bernoulli shows that the cycloid is the solution to the tautochrone problem

- 1691 - Johann Bernoulli shows that a chain freely suspended from two points will form a catenary

- 1691 - James Bernoulli shows that the catenary curve has the lowest center of gravity that any chain hung from two fixed points can have

- 1696 - Johann Bernoulli shows that the cycloid is the solution to the brachistochrone problem

- 1714 - Brook Taylor derives the fundamental frequency of a stretched vibrating string in terms of its tension and mass per unit length by solving an ordinary differential equation

- 1733 - Daniel Bernoulli derives the fundamental frequency and harmonics of a hanging chain by solving an ordinary differential equation

- 1734 - Daniel Bernoulli solves the ordinary differential equation for the vibrations of an elastic bar clamped at one end

- 1738 - Daniel Bernoulli examines fluid flow in Hydrodynamica

- 1739 - Leonhard Euler solves the ordinary differential equation for a forced harmonic oscillator and notices the resonance phenomenon

- 1742 - Colin Maclaurin discovers his uniformly rotating self-gravitating spheroids

- 1743 - Jean le Rond d'Alembert publishes his "Traite de Dynamique", in which he introduces the concept of generalized forces for accelerating systems and systems with constraints

- 1747 - Pierre Louis Maupertuis applies minimum principles to mechanics

- 1759 - Leonhard Euler solves the partial differential equation for the vibration of a rectangular drum

- 1764 - Leonhard Euler examines the partial differential equation for the vibration of a circular drum and finds one of the Bessel function solutions

- 1776 - John Smeaton publishes a paper on experiments relating power, work, momentum and kinetic energy, and supporting the conservation of energy

- 1788 - Joseph Louis Lagrange presents Lagrange's equations of motion in *Mécanique Analytique*

- 1789 - Antoine Lavoisier states the law of conservation of mass

- 1813 - Peter Ewart supports the idea of the conservation of energy in his paper *On the measure of moving force*

- 1821 - William Hamilton begins his analysis of Hamilton's characteristic function

- 1834 - Carl Jacobi discovers his uniformly rotating self-gravitating ellipsoids

- 1834 - John Russell observes a nondecaying solitary water wave (soliton) in the Union Canal near Edinburgh and uses a water tank to study the dependence of solitary water wave velocities on wave amplitude and water depth

- 1835 - William Hamilton states Hamilton's canonical equations of motion

- 1835 - Gaspard Coriolis examines theoretically the mechanical efficiency of waterwheels, and deduces the Coriolis effect.

- 1841 - Julius Robert von Mayer, an amateur scientist, writes a paper on the conservation of energy but his lack of academic training leads to its rejection.

- 1842 - Christian Doppler proposes the Doppler effect

- 1847 - Hermann von Helmholtz formally states the law of conservation of energy

- 1851 - Léon Foucault shows the Earth's rotation with a huge pendulum (Foucault pendulum)

- 1902 - James Jeans finds the length scale required for gravitational perturbations to grow in a static nearly homogeneous medium

4.3   References

[1] O'Connor, John J.; Robertson, Edmund F., "Al-Biruni", *MacTutor History of Mathematics archive*, University of St Andrews.:

> "One of the most important of al-Biruni's many texts is *Shadows* which he is thought to have written around 1021. [...] *Shadows* is an extremely important source for our knowledge of the history of mathematics, astronomy, and physics. It also contains important ideas such as the idea that acceleration is connected with non-uniform motion, using three rectangular coordinates to define a point in 3-space, and ideas that some see as anticipating the introduction of polar coordinates."

[2] Shlomo Pines (1964), "La dynamique d'Ibn Bajja", in *Mélanges Alexandre Koyré*, I, 442-468 [462, 468], Paris.
(cf. Abel B. Franco (October 2003). "Avempace, Projectile Motion, and Impetus Theory", *Journal of the History of Ideas* **64** (4), p. 521-546 [543]: *"Pines has also seen Avempace's idea of fatigue as a precursor to the Leibnizian idea of force which, according to him, underlies Newton's third law of motion and the concept of the "reaction" of forces."*)

[3] Pines, Shlomo (1970). "Abu'l-Barakāt al-Baghdādī , Hibat Allah". *Dictionary of Scientific Biography* **1**. New York: Charles Scribner's Sons. pp. 26–28. ISBN 0-684-10114-9.:
(cf. Abel B. Franco (October 2003). "Avempace, Projectile Motion, and Impetus Theory", *Journal of the History of Ideas* **64** (4), p. 521-546 [528]: *Hibat Allah Abu'l-Barakat al-Bagdadi (c.1080- after 1164/65) extrapolated the theory for the case of falling bodies in an original way in his Kitab al-Mu'tabar (The Book of that Which is Established through Personal Reflection). [...] This idea is, according to Pines, "the oldest negation of Aristotle's fundamental dynamic law [namely, that a constant force produces a uniform motion]," and is thus an "anticipation in a vague fashion of the fundamental law of classical mechanics [namely, that a force applied continuously produces acceleration]."*)

[4] Mariam Rozhanskaya and I. S. Levinova (1996), "Statics", in Roshdi Rashed, ed., *Encyclopedia of the History of Arabic Science*, Vol. 2, p. 614-642 [621], Routledge, London and New York

[5] F. Jamil Ragep (2001), "Tusi and Copernicus: The Earth's Motion in Context", *Science in Context* **14** (1-2), p. 145–163. Cambridge University Press.

Chapter 5

Timeline of electromagnetism and classical optics

Timeline of electromagnetism and classical optics

5.1 Early developments

- 424 BC Aristophanes "lens" is a glass globe filled with water.(Seneca says that it can be used to read letters *no matter how small or dim*)[1]

- 4th century BC Mo Di first mentions the camera obscura, a pin-hole camera.

- 3rd century BC Euclid is the first to write about reflection and refraction and notes that light travels in straight lines[1]

- 130 AD. — Claudius Ptolemy (in his work *Optics*) wrote about the properties of light including: reflection, refraction, and color and tabulated angles of refraction for several media

- 1021 — Ibn al-Haytham (Alhazen) writes the *Book of Optics*, studying vision.

- 1088 — Shen Kuo first recognizes magnetic declination.

- 1187 — Alexander Neckham is first in Europe to describe the magnetic compass and its use in navigation.

- 1269 — Pierre de Maricourt describes magnetic poles and remarks on the nonexistence of isolated magnetic poles

- 1305 — Dietrich von Freiberg uses crystalline spheres and flasks filled with water to study the reflection and refraction in raindrops that leads to primary and secondary rainbows

- 1550 — Gerolamo Cardano writes about electricity in *De Subtilitate* distinguishing, perhaps for the first time, between electrical and magnetic forces.

5.2 17th century

- 1600 — Dutchman Sacharias Jansen invents a single-lens microscope.

- 1600 — William Gilbert, in his book de Magnete, wrote about systematic experiments in electricity and magnetism; deduced that the Earth is a giant magnet.

- 1604 — Johannes Kepler describes how the eye focuses light

- 1604 — Johannes Kepler specifies the laws of the rectilinear propagation of light

- 1611 — Marko Dominis discusses the rainbow in *De Radiis Visus et Lucis*

- 1611 — Johannes Kepler discovers total internal reflection, a small-angle refraction law, and thin lens optics,

- 1621 — Willebrord van Roijen Snell states his Snell's law of refraction

- 1630 — Cabaeus finds that there are two types of electric charges

- 1637 — René Descartes quantitatively derives the angles at which primary and secondary rainbows are seen with respect to the angle of the Sun's elevation

- 1646 — Sir Thomas Browne first uses the word *electricity* is in his work Pseudodoxia Epidemica.

- 1657 — Pierre de Fermat introduces the principle of least time into optics

- 1660 — Otto von Guericke invents an early electrostatic generator.

- 1665 — Francesco Maria Grimaldi highlights the phenomenon of diffraction

- 1673 — Ignace Pardies provides a wave explanation for refraction of light

- 1675 — Robert Boyle discovers that electric attraction and repulsion can act across a vacuum and do not depend upon the air as a medium. Adds resin to the known list of "electrics."

- 1675 — Isaac Newton delivers his theory of light

- 1676 — Olaus Roemer measures the speed of light by observing Jupiter's moons

- 1678 — Christiaan Huygens states his principle of wavefront sources and demonstrates the refraction and diffraction of light rays.

5.3   18th century

- 1704 — Isaac Newton publishes *Opticks*, a corpuscular theory of light and colour

- 1728 — James Bradley discovers the aberration of starlight and uses it to determine that the speed of light is about 283,000 km/s

- 1729 — Stephen Gray demonstrates the difference between conductors and non-conductors (insulators).

- 1732 — C. F. du Fay Shows that all objects, except metals, animals, and liquids, can be electrified by rubbing them and that metals, animals and liquids could be electrified by means of an electrostatic generators

- 1737 — C. F. du Fay and Francis Hauksbee the younger independently discover two kinds of frictional electricity: one generated from rubbing glass, the other from rubbing resin (later identified as positive and negative electrical charges).

- 1740 — Jean le Rond d'Alembert, in *Mémoire sur la réfraction des corps solides*, explains the process of refraction.

- 1745 — Pieter van Musschenbroek invents the Leyden jar, a type of capacitor.

- 1746 — Leonhard Euler develops the wave theory of light refraction and dispersion

- 1747 — William Watson, while experimenting with a Leyden jar, observes that a discharge of static electricity causes electric current to flow and develops the concept of an electrical potential (voltage).

- 1752 — Benjamin Franklin shows that lightning is electricity. Also credited with the convention of using "negative" and "positive" to denote an electrical charge or potential.

- 1767 — Joseph Priestley proposes an electrical inverse-square law

- 1784 — Henry Cavendish defines the inductive capacity of dielectrics (insulators) and measures the specific inductive capacity of various substances by comparison with an air condenser.

- 1785 — Charles Coulomb introduces the inverse-square law of electrostatics

- 1786 — Luigi Galvani discovers "animal electricity" and postulates that animal bodies are storehouses of electricity. His invention of the voltaic cell leads to the invention the electric battery.

5.4 19th century

5.4.1 1800-1850

- 1800 — William Herschel discovers infrared radiation from the Sun

- 1800 — William Nicholson and Johann Ritter use electricity to decompose water into hydrogen and oxygen, thereby discovering the process of electrolysis, which led to the discovery of many other elements.

- 1800 — Alessandro Volta invents the voltaic pile, or "battery", specifically to disprove Galvani's animal electricity theory.

- 1801 — Johann Ritter discovers ultraviolet radiation from the Sun

- 1801 — Thomas Young demonstrates the wave nature of light and the principle of interference

- 1802 — Gian Domenico Romagnosi notes that a nearby voltaic pile deflects a magnetic needle. His account is largely overlooked.

- 1803 — Thomas Young develops the Double-slit experiment and demonstrates the effect of interference.

- 1806 — Alessandro Volta employs a voltaic pile to decompose potash and soda, showing that they are the oxides of the previously unknown metals potassium and sodium. These experiments were the beginning of electrochemistry.

- 1808 — Étienne-Louis Malus discovers polarization by reflection

- 1809 — Étienne-Louis Malus publishes the law of Malus which predicts the light intensity transmitted by two polarizing sheets

- 1809 — Humphry Davy first publicly demonstrates the electric arc light.

- 1811 — François Jean Dominique Arago discovers that some quartz crystals continuously rotate the electric vector of light

- 1816 — David Brewster discovers stress birefringence

- 1818 — Siméon Poisson predicts the Poisson-Arago bright spot at the center of the shadow of a circular opaque obstacle

- 1818 — François Jean Dominique Arago verifies the existence of the Poisson-Arago bright spot

- 1820 — Hans Christian Ørsted notices that a current in a wire can deflect a compass needle, demonstrating a relationship between electricity and magnetism

- 1821 — André-Marie Ampère announces his theory of electrodynamics, predicting the force that one current exerts upon another.

- 1821 — Thomas Johann Seebeck discovers the thermoelectric effect.

- 1821 — Augustin-Jean Fresnel derives a mathematical demonstration that polarization can be explained only if light is *entirely* transverse, with no longitudinal vibration whatsoever.

- 1825 — Augustin Fresnel phenomenologically explains optical activity by introducing circular birefringence

- 1826 — Georg Simon Ohm states his Ohm's law of electrical resistance

- 1831 — Michael Faraday states his law of induction

- 1831 — Macedonio Melloni uses a thermopile to detect infrared radiation

- 1833 — Heinrich Lenz states that an induced current in a closed conducting loop will appear in such a direction that it opposes the change that produced it (Lenz's law)

- 1833 — Michael Faraday announces his law of electrochemical equivalents

- 1834 — Heinrich Lenz determines the direction of the induced electromotive force (emf) and current resulting from electromagnetic induction. Lenz's law provides a physical interpretation of the choice of sign in Faraday's law of induction (1831), indicating that the induced emf and the change in flux have opposite signs.

- 1834 — Jean-Charles Peltier discovers the Peltier effect: heating by an electric current at the junction of two different metals.

- 1838 — Michael Faraday uses Volta's battery to discover cathode rays.

- 1839 — Alexandre Edmond Becquerel observes the photoelectric effect with an electrode in a conductive solution exposed to light.

- 1845 — Michael Faraday discovers that light propagation in a material can be influenced by external magnetic fields (Faraday effect)

- 1849 — Hippolyte Fizeau and Jean-Bernard Foucault measure the speed of light to be about 298,000 km/s

5.4.2 1851-1899

- 1852 — George Gabriel Stokes defines the Stokes parameters of polarization

- 1852 — Edward Frankland develops the theory of chemical valence

- 1864 — James Clerk Maxwell publishes his papers on a dynamical theory of the electromagnetic field

- 1869 — William Crookes invents the Crookes tube.

- 1873 — Willoughby Smith discovers the photoelectric effect in metals not in solution (i.e., selenium).

- 1871 — Lord Rayleigh discusses the blue sky law and sunsets (Rayleigh scattering)

- 1873 — James Clerk Maxwell states that light is an electromagnetic phenomenon

- 1875 — John Kerr discovers the electrically induced birefringence of some liquids

- 1879 — Jožef Stefan discovers the Stefan-Boltzmann radiation law of a black body and uses it to calculate the first sensible value of the temperature of the Sun's surface to be 5700 K

- 1886 — Oliver Heaviside coins the term *inductance*.

- 1887 — Heinrich Hertz invents a device for the production and reception of electromagnetic (EM) radio waves. His receiver consists of a coil with a spark gap.

- 1888 — Heinrich Rudolf Hertz discovers radio waves

- 1893 — Victor Schumann discovers the vacuum ultraviolet spectrum.

- 1895 — Wilhelm Conrad Röntgen discovers X-rays

- 1895 — Jagadis Chandra Bose gives his first public demonstration of electromagnetic waves

- 1896 — Arnold Sommerfeld solves the half-plane diffraction problem

5.5   20th century

- 1900 — The Liénard–Wiechert potentials are introduced as time-dependent (retarded) electrodynamic potentials

- 1905 — Albert Einstein demonstrates that Maxwell's Equations are not required to describe electromagnetic radiation if Special Relativity is taken into account

- 1919 — Albert A. Michelson makes the first interferometric measurements of stellar diameters at Mount Wilson Observatory (see history of astronomical interferometry)

- 1946 — Martin Ryle and Vonberg build the first two-element astronomical radio interferometer (see history of astronomical interferometry)

- 1953 — Charles H. Townes, James P. Gordon, and Herbert J. Zeiger produce the first maser

- 1956 — R. Hanbury-Brown and R.Q. Twiss complete the correlation interferometer

- 1960 — Theodore Maiman produces the first working laser

- 1966 — Jefimenko introduces time-dependent (retarded) generalizations of Coulomb's law and the Biot-Savart law

- 1999 — M. Henny and others demonstrate the Fermionic Hanbury Brown and Twiss Experiment

5.6   Notes

[1] *The history of the telescope* by Henry C. King, Harold Spencer Jones Publisher Courier Dover Publications, 2003 Pg 25 ISBN 0-486-43265-3, ISBN 978-0-486-43265-6

5.7   External links

- The Work of Jagadis Chandra Bose: 100 Years of MM-Wave Research

- Jagadis Chandra Bose and His Pioneering Research on Microwaves

Chapter 6

Timeline of electromagnetic theory

The **timeline of electromagnetism**, that is the timeline of the human understanding of electromagnetic forces, dates back over two thousand years ago. It lists, within the history of electromagnetism, the associated theories, technology, and events.

6.1 Ancient history

6th century BC: Thales of Miletus is credited with observing that rubbing fur on various substances, such as amber, would cause an attraction between the two, which is now known to be caused by static electricity. The Ancient Greeks noted that the amber buttons could attract light objects such as hair and that if the amber was rubbed sufficiently a spark would jump.

3rd century BC: the Baghdad Battery is dated from this period. It resembles a galvanic cell and is believed by some to have been used for electroplating, although there is no common consensus on the purpose of these devices nor whether they were, indeed, even electrical in nature.[1]

1st century BC: Pliny in his Natural History records the story of a shepherd Magnes who discovered the magnetic properties of some iron stones, *"it is said, made this discovery, when, upon taking his herds to pasture, he found that the nails of his shoes and the iron ferrel of his staff adhered to the ground."*[2]

6.2 The Renaissance

1550: Girolamo Cardano distinguishes between electrical and magnetic forces in *De subtilitate rerum*.

1600: William Gilbert publishes *De Magnete, Magneticisque Corporibus, et de Magno Magnete Tellure* ("On the Magnet and Magnetic bodies, and on that Great Magnet the Earth"), Europe's then current standard on electricity and magnetism. He experimented with and noted the different character of electrical and magnetic forces. In addition to known ancient Greeks' observations of the electrical properties of rubbed amber, he experimented with a needle balanced on a pivot, and found that the needle was non-directionally affected by many materials such as alum, arsenic, hard resin, jet, glass, gum-mastic, mica, rock-salt, sealing wax, slags, sulfur, and precious stones such as amethyst, beryl, diamond, opal, and sapphire. He noted that electrical charge could be stored by covering the body with a non-conducting substance such as silk. He described the method of artificially magnetizing iron. His terrella (little earth), a sphere cut from a lodestone on a metal lathe, modeled the earth as a lodestone (magnetic iron ore) and demonstrated that every lodestone has fixed poles, and how to find them.[3] He considered that gravity was a magnetic force and noted that this mutual force increased with the size or amount of lodestone and attracted iron objects. He experimented with such physical models in an attempt to explain problems in navigation due varying properties of the magnetic compass with respect to their location on the earth, such as magnetic declination and magnetic inclination. His experiments explained the dipping of the needle by the magnetic attraction of the earth, and were used to predict where the vertical dip would be found. Such magnetic

Detail of the right-hand facade fresco, showing Thales of Miletus, National and Kapodistrian University of Athens.

inclination was described as early as the 11th century by Shen Kuo in his *Meng Xi Bi Tan* and further investigated in 1581 by retired mariner and compass maker Robert Norman, as described in his pamphlet, *The Newe Attractive.* The gilbert, a unit of magnetomotive force or magnetic potential, was named in his honor.

1646: Sir Thomas Browne uses the word electricity in *Pseudodoxia Epidemica.*

1663: Otto von Guericke (brewer and engineer who applied the barometer to weather prediction and invented the air pump, with which he demonstrated the properties of atmospheric pressure associated with a vacuum) constructs a primitive electrostatic generating (or friction) machine via the triboelectric effect, utilizing a continuously rotating sulfur globe that could be rubbed by hand or a piece of cloth. Isaac Newton suggested the use of a glass globe instead of a sulfur one.

1675: Robert Boyle states that electric attraction and repulsion can act across a vacuum.

1705: Francis Hauksbee improves von Guericke's electrostatic generator by using a glass globe and generates the first sparks by approaching his finger to the rubbed globe.

1729: Stephen Gray and the Reverend Granville Wheler experiment to discover that electrical "virtue," produced by rubbing a glass tube, could be transmitted over an extended distance (nearly 900 ft (about 270 m)) through thin iron wire using silk threads as insulators, to deflect leaves of brass. This has been described as the beginning of electrical communication.[4] This was also the first distinction between the roles of conductors and insulators (names applied by John Desaguliers, mathematician and Royal Society member, who stated that Gray "has made greater variety of electrical experiments than all the philosophers of this and the last age.")[4] Georges-Louis LeSage built a static electricity telegraph in 1774, based upon the same principles discovered by Gray.

1734: Charles François de Cisternay DuFay (inspired by Gray's work to perform electrical experiments) dispels the effluvia theory by his paper in Volume 38 of the *Philosophical Transactions of the Royal Society,* describing his discovery of the distinction between two kinds of electricity: "resinous," produced by rubbing bodies such as amber, copal, or gum-lac with silk or paper, and "vitreous," by rubbing bodies as glass, rock crystal, or precious stones with hair or wool. He also posited the principle of mutual attraction for unlike forms and the repelling of like forms and that "from this principle one may with ease deduce the explanation of a great number of other phenomena." The terms resinous and vitreous were later replaced with the terms "positive" and "negative" by William Watson and Benjamin Franklin.

1745: Pieter van Musschenbroek of Leiden (Leyden) independently discovers the Leyden (Leiden) jar, a primitive capacitor or "condenser" (term coined by Volta in 1782, derived from the Italian *condensatore*), with which the transient electrical energy generated by current friction machines could now be stored. He and his student Andreas Cunaeus used a glass jar filled with water into which a brass rod had been placed. He charged the jar by touching a wire leading from the electrical machine with one hand while holding the outside of the jar with the other. The energy could be discharged by completing an external circuit between the brass rod and another conductor, originally his hand, placed in contact with the outside of the jar. He also found that if the jar were placed on a piece of metal on a table, a shock would be received by touching this piece of metal with one hand and touching the wire connected to the electrical machine with the other.

1745: Ewald Georg von Kleist of independently invents the capacitor: a glass jar coated inside and out with metal. The inner coating was connected to a rod that passed through the lid and ended in a metal sphere. By having this thin layer of glass insulation (a dielectric) between two large, closely spaced plates, von Kleist found the energy density could be increased dramatically compared with the situation with no insulator. Daniel Gralath improved the design and was also the first to combine several jars to form a battery strong enough to kill birds and small animals upon discharge.

1752: Benjamin Franklin establishes the link between lightning and electricity by the flying a kite into a thunderstorm and transferring some of the charge into a Leyden jar and showed that its properties were the same as charge produced by an electrical machine. He is credited with utilizing the concepts of positive and negative charge in the explanation of then known electrical phenomenon. He theorized that there was an electrical fluid (which he proposed could be the luminiferous ether, which was used by others before and after him, to explain the wave theory of light) that was part of all material and all intervening space. The charge of any object would be neutral if the concentration of this fluid were the same both inside and outside of the body, positive if the object contained an excess of this fluid, and negative if there were a deficit. In 1749 he had documented the similar properties of lightning and electricity, such as that both an electric spark and a lightning flash produced light and sound, could kill animals, cause fires, melt metal, destroy or reverse the polarity of magnetism, and flowed through conductors and could be concentrated at sharp points. He was later able to apply the property of concentrating at sharp points by his invention of the lightning rod, for which he intentionally did not profit. He also investigated the Leyden jar, proving that the charge was stored on the glass and not in the water, as others had assumed.

1753: C. M. (of Scotland, possibly Charles Morrison, of Greenock or Charles Marshall, of Aberdeen) proposes in the 17 February edition of Scots Magazine, an electrostatic telegraph system with 26 insulated wires, each corresponding to a letter of the alphabet and each connected to electrostatic machines. The receiving charged end was to electrostatically attract a disc of paper marked with the corresponding letter.

1767: Joseph Priestley proposes an electrical inverse-square law.

1774: Georges-Louis LeSage builds an electrostatic telegraph system with 26 insulated wires conducting Leyden-jar charges to pith-ball electroscopes, each corresponding to a letter of the alphabet. Its range was only between rooms of his home.

1785: Charles Coulomb introduces the inverse-square law of electrostatics.

1791: Luigi Galvani discovers galvanic electricity and bioelectricity through experiments following an observation that touching exposed muscles in frogs' legs with a scalpel which had been close to a static electrical machine caused them to jump. He called this "animal electricity". Years of experimentation in the 1780s eventually led him to the construction of an arc of two different metals (copper and zinc for example) by connecting the two metal pieces and then connecting their open ends across the nerve of a frog leg, producing the same muscular contractions (by completing a circuit) as originally accidentally observed. The use of different metals to produce an electrical spark is the basis that led Alessandro Volta in 1799 to his invention of his voltaic pile, which eventually became the galvanic battery.[5]

1799: Alessandro Volta, following Galvani's discovery of galvanic electricity, creates a voltaic cell producing an electric current by the chemical action of several pairs of alternating copper (or silver) and zinc discs "piled" and separated by cloth or cardboard which had been soaked brine (salt water) or acid to increase conductivity. In **1800** he demonstrates the production of light from a glowing wire conducting electricity. This was followed in 1801 by his construction of the first electric battery, by utilizing multiple voltaic cells. Prior to his major discoveries, in a letter of praise to the Royal Society 1793, Volta reported Luigi Galvani's experiments of the 1780s as the "most beautiful and important discoveries," regarding them as the foundation of future discoveries. Volta's inventions led to revolutionary changes with this method of the production of inexpensive, controlled electric current vs. existing frictional machines and Leyden jars. The electric battery became standard equipment in every experimental laboratory and heralded an age of practical applications of electricity.[4] The unit volt is named for his contributions.

1800: William Nicholson and Anthony Carlisle discover electrolysis by passing a voltaic current through water, decomposing it into its elements hydrogen and oxygen.

1802: Gian Domenico Romagnosi, Italian legal scholar, discovers that electricity and magnetism are related by noting that a nearby voltaic pile deflects a magnetic needle. He published his account in an Italian newspaper, but this was overlooked by the scientific community.[6]

1820: Hans Christian Ørsted, Danish physicist and chemist, unites the separate sciences of electricity and magnetism. He develops an experiment in which he notices a compass needle is deflected from magnetic north when an electric current from the battery he was using was switched on and off, convincing him that magnetic fields radiate from all sides of a live wire just as light and heat do, confirming a direct relationship between electricity and magnetism. He also observes that the movement of the compass-needle to one side or the other depends upon the direction of the current. Following intensive investigations, he published his findings, proving that a changing electric current produces a magnetic field as it flows through a wire. The oersted unit of magnetic induction is named for his contributions.

1820: André-Marie Ampère, professor of mathematics at the Ecole Polytechnique, a short time after learning of Ørsted's discovery that magnetic needle is acted on by a voltaic current, conducts experiments and publishes a paper in *Annales de Chimie et de Physique* attempting to give a combined theory of electricity and magnetism. He shows that a coil of wire carrying a current behaves like an ordinary magnet and suggests that electromagnetism might be used in telegraphy. He mathematically develops Ampère's law describing the magnetic force between two electric currents. His mathematical theory explains known electromagnetic phenomena and predicts new ones. His laws of electrodynamics include the facts that parallel conductors currying current in the same direction attract and those carrying currents in the opposite directions repel one another. One of the first to develop electrical measuring techniques, he built an instrument utilizing a free-moving needle to measure the flow of electricity, contributing to the development of the galvanometer. In **1821**, he proposed a telegraphy system utilizing one wire per "galvanometer" to indicate each letter, and reported experimenting successfully with such a system. However, in **1824**, Peter Barlow reported its maximum distance was only 200 feet, and so was impractical. In **1826** he publishes the *Memoir on the Mathematical Theory of Electrodynamic Phenomena, Uniquely Deduced from Experience* containing a mathematical derivation of the electrodynamic force law. Following Faraday's discovery of electromagnetic induction in 1831, Ampère agreed that Faraday deserved full credit for the discovery.

1820: Johann Salomo Christoph Schweigger, German chemist, physicist, and professor, builds the first sensitive galvanometer, wrapping a coil of wire around a graduated compass, an acceptable instrument for actual measurement as well as detection of small amounts of electric current, naming it after Luigi Galvani.

~1825: William Sturgeon, founder of the first English Electric Journal, *Annals of Electricity*, found that an iron core inside a helical coil of wire connected to a battery greatly increased the resulting magnetic field, thus making possible the more powerful electromagnets utilizing a ferromagnetic core. Sturgeon also bent the iron core into a U-shape to bring the poles closer together, thus concentrating the magnetic field lines. These discoveries followed Ampère's discovery that electricity passing through a coiled wire produced a magnetic force and that of Dominique François Jean Arago finding that an iron bar is magnetized by putting it inside the coil of current-carrying wire, but Arago had not observed the increased strength of the resulting field while the bar was being magnetized.

1826: Georg Simon Ohm states his Ohm's law of electrical resistance in the journals of Schweigger and Poggendorff, and also published in his landmark pamphlet *Die galvanische Kette mathematisch bearbeitet* in **1827**. The unit ohm (Ω) of electrical resistance has been named in his honor.[7]

1829 & 1830: Francesco Zantedeschi publishes papers on the production of electric currents in closed circuits by the approach and withdrawal of a magnet, thereby anticipating Michael Faraday's classical experiments of 1831.

6.3   Modern Developments

1831: Michael Faraday began experiments leading to his discovery of the law of electromagnetic induction, though the discovery may have been anticipated by the work of Francesco Zantedeschi. His breakthrough came when he wrapped two insulated coils of wire around a massive iron ring, bolted to a chair, and found that upon passing a current through one coil, a momentary electric current was induced in the other coil. He then found that if he moved a magnet through a loop of wire, or vice versa, an electric current also flowed in the wire. He then used this principle to construct the electric dynamo, the first electric power generator. He proposed that electromagnetic forces extended into the empty space around the conductor, but did not complete that work. Faraday's concept of lines of flux emanating from charged bodies and magnets provided a way to visualize electric and magnetic fields. That mental model was crucial to the successful development of electromechanical devices which were to dominate the 19th century. His demonstrations that a changing magnetic field produces an electric field, mathematically modeled by Faraday's law of induction, would subsequently become one of Maxwell's equations. These consequently evolved into the generalization of field theory.

1832: Baron Pavel L'vovitch Schilling (Paul Schilling) creates the first electromagnetic telegraph, consisting of a single-needle system in which a code was used to indicate the characters. Only months later, Göttingen professors Carl Friedrich Gauss and Wilhelm Weber constructed a telegraph that was working two years before Schilling could put his into practice. Schilling demonstrated the long-distance transmission of signals between two different rooms of his apartment and was the first to put into practice a binary system of signal transmission.

1833: Heinrich Lenz states Lenz's law: if an increasing (or decreasing) magnetic flux induces an electromotive force (EMF), the resulting current will oppose a further increase (or decrease) in magnetic flux, i.e., that an induced current in a closed conducting loop will appear in such a direction that it opposes the change that produced it. Lenz's law is one consequence of the principle of conservation of energy. If a magnet moves towards a closed loop, then the induced current in the loop creates a field that exerts a force opposing the motion of the magnet. Lenz's law can be derived from Faraday's law of induction by noting the negative sign on the right side of the equation. He also independently discovered Joule's law in **1842**; to honor his efforts, Russian physicists refer to it as the "Joule-Lenz law."

1835: Joseph Henry invents the electric relay, which is an electrical switch by which the change of a weak current through the windings of an electromagnet will attract an armature to open or close the switch. Because this can control (by opening or closing) another, much higher-power, circuit, it is in a broad sense a form of electrical amplifier. This made a practical electric telegraph possible. He was the first to coil insulated wire tightly around an iron core in order to make an extremely powerful electromagnet, improving on William Sturgeon's design, which used loosely coiled, uninsulated wire. He also discovered the property of self inductance independently of Michael Faraday.

International Morse Code

1. The length of a dot is one unit.
2. A dash is three units.
3. The space between parts of the same letter is one unit.
4. The space between letters is three units.
5. The space between words is seven units.

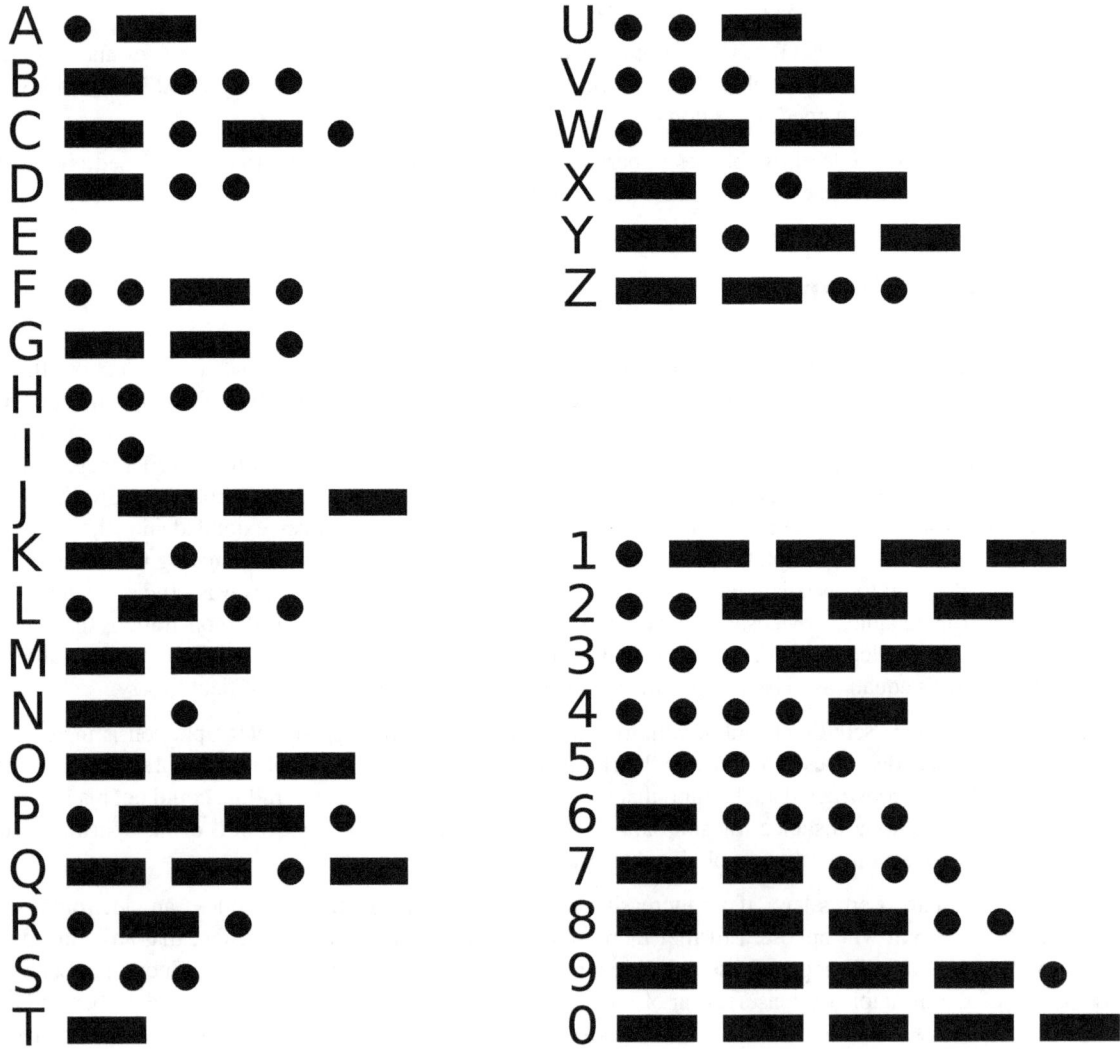

Chart of the Morse code letters and numerals.

1836: Dr. David Alter invents and demonstrates to witnesses the first American electric telegraph in Elderton, Pennsylvania. In a later interview in the book, *Biographical and Historical Cyclopedia of Indiana and Armstrong Counties* he states: "I may say that there is no connection at all between the telegraph of Morse and others and that of myself...Professor Morse most probably never heard of me or my Elderton telegraph." In **1840** he invents an electric buggy, forerunner of the automobile. His inventions also include an electric clock and a short-range type of telephone, forerunner to Alexander Graham Bell's telephone. He is also credited with the origins of Spectrum Analysis by his idea that every element has its own emission spectrum, and an expansion of spectrum analysis to include the optical properties of gases.

1837: Samuel Morse develops an alternative electrical telegraph design capable of transmitting long distances over poor quality wire. He and his assistant Alfred Vail develop the Morse code signaling alphabet. In **1838** Morse successfully tested the device at the Speedwell Ironworks near Morristown, New Jersey, and publicly demonstrated it to a scientific committee at the Franklin Institute in Philadelphia, Pennsylvania. The first electric telegram using this device was sent by Morse on 24 May, **1844** from Baltimore to Washington, D.C., bearing the message "What hath God wrought?"

1840: James Prescott Joule formulates Joule's Law (sometimes called the Joule-Lenz law) quantifying the amount of heat produced in a circuit as proportional to the product of the time duration, the resistance, and the square of the current passing through it.

1845: Michael Faraday discovers that light propagation in a material can be influenced by external magnetic fields.

1849: Hippolyte Fizeau and Jean-Bernard Foucault measure the speed of light to be about 298,000 km/s.

1854: Gustav Robert Kirchhoff, physicist and one of the founders of spectroscopy, publishes Kirchhoff's Laws on the conservation of electric charge and energy, which are used to determine currents in each branch of a circuit.

1861: the first transcontinental telegraph system spans North America by connecting an existing network in the eastern United States to a small network in California by a link between Omaha and Carson City via Salt Lake City. The slower Pony Express system ceased operation a month later.

1865: James Clerk Maxwell publishes his landmark paper A Dynamical Theory of the Electromagnetic Field, in which Maxwell's equations demonstrated that electric and magnetic forces are two complementary aspects of electromagnetism. He shows that the associated complementary electric and magnetic fields of electromagnetism travel through space, in the form of waves, at a constant velocity of 3.0×10^8 m/s. He also proposes that light is a form of electromagnetic radiation and that waves of oscillating electric and magnetic fields travel through empty space at a speed that could be predicted from simple electrical experiments. Using available data, he obtains a velocity of 310,740,000 m/s and states "This velocity is so nearly that of light, that it seems we have strong reason to conclude that light itself (including radiant heat, and other radiations if any) is an electromagnetic disturbance in the form of waves propagated through the electromagnetic field according to electromagnetic laws."

1866: the first successful **transatlantic telegraph system** was completed. Earlier submarine cable transatlantic cables installed in 1857 and 1858 failed after operating for a few days or weeks.

1873: J. C. Maxwell publishes A Treatise on Electricity and Magnetism which states that light is an electromagnetic phenomenon.

1874: German scientist Karl Ferdinand Braun discovers the "unilateral conduction" of crystals.[8][9] Braun patents the first solid state diode, a crystal rectifier, in **1899**.[10]

1878: Thomas Edison, following work on a "multiplex telegraph" system and the phonograph, invents an improved incandescent light bulb. This was not the first electric light bulb but the first commercially practical incandescent light. In **1879** he produces a high-resistance lamp in a very high vacuum; the lamp lasts hundreds of hours. While the earlier inventors had produced electric lighting in lab conditions, Edison concentrated on commercial application and was able to sell the concept to homes and businesses by mass-producing relatively long-lasting light bulbs and creating a complete system for the generation and distribution of electricity.

1880: Edison discovers thermionic emission or the Edison effect.

1882: Edison switches on the world's first electrical power distribution system, providing 110 volts direct current (DC) to 59 customers.

1884: Oliver Heaviside reformulates Maxwell's original mathematical treatment of electromagnetic theory from twenty equations in twenty unknowns into four simple equations in four unknowns (the modern vector form of Maxwell's equations).

1887: Nikola Tesla develops an induction motor that uses alternating current, or AC, instead of direct current.

1888: Heinrich Hertz demonstrates the existence of electromagnetic waves by building an apparatus that produced and detected UHF radio waves (or microwaves in the UHF region). He also found that radio waves could be transmitted through different types of materials and were reflected by others, the key to radar. His experiments explain reflection, refraction, polarization, interference, and velocity of electromagnetic waves.

1897: J. J. Thomson discovers the electron.

1900: Max Planck resolves the ultraviolet catastrophe by suggesting that black body radiation consists of discrete packets, or quanta, of energy. The amount of energy in each packet is proportional to the frequency of the electromagnetic waves. The constant of proportionality is now called Planck's constant in his honor.

1904: John Ambrose Fleming invents the thermionic diode, the first electronic vacuum tube, which had practical use in early radio receivers.

1905: Albert Einstein proposes the Theory of Special Relativity, in which he rejects the existence of the aether as unnecessary for explaining the propagation of electromagnetic waves. Instead, Einstein asserts *as a postulate* that the speed of light is constant in all inertial frames of reference, and goes on to demonstrate a number of revolutionary (and highly counter-intuitive) consequences, including time dilation, length contraction, the relativity of simultaneity, the dependence of mass on velocity, and the equivalence of mass and energy.

1905: Einstein explains the photoelectric effect by extending Planck's idea of light quanta, or photons, to the absorption and emission of photoelectrons. Einstein would later receive the Nobel Prize in Physics for this discovery, which launched the quantum revolution in physics.

1911: Superconductivity is discovered by Heike Kamerlingh Onnes, who was studying the resistivity of solid mercury at cryogenic temperatures using the recently discovered liquid helium as a refrigerant. At the temperature of 4.2 K, he observed that the resistivity abruptly disappeared. For this discovery, he was awarded the Nobel Prize in Physics in 1913.

1924: Louis de Broglie postulates the wave nature of electrons and suggests that all matter has wave properties.

6.4   See also

- History of electromagnetic theory
- History of special relativity
- History of superconductivity
- Timeline of luminiferous aether

6.5   References

[1] Frood, Arran (27 February 2003). "Riddle of 'Baghdad's batteries'". *BBC News*. Retrieved 20 October 2015.

[2] Pliny the Elder. "Dedication". *The Natural History*. Perseus Collection: Greek and Roman Materials. Department of the Classics, Tufts University. Retrieved 20 October 2015.

[3] Williams, Henry Smith. "Part IV. William Gilbert and the Study of Magnetism". *A history of science* **2**. Worldwide School. Retrieved 20 October 2015.

[4] Clark, David H.; Clark, Stephen P.H. (2001). *Newton's tyranny : the suppressed scientific discoveries of Stephen Gray and John Flamsteed*. New York: Freeman. ISBN 9780716747017.

[5] Williams, Henry Smith. "VII. The Modern Development of Electricity and Magnetism". *A history of science* **3**. Worldwide School. Retrieved 20 October 2015.

[6] Martins, Roberto de Andrade. "Romagnosi and Volta's pile: early difficulties in the interpretation of Voltaic electricity". In Bevilacqua, Fabio; Fregonese, Lucio. *Nuova Voltiana: Studies on Volta and his Times* **3**. Pavia: Ulrico Hoepli. pp. 81–102.

[7] "Georg Simon Ohm: The Discovery of Ohm's Law". Juliantrubin.com. Retrieved 2011-11-15.

[8] Braun, Ferdinand (1874) "Ueber die Stromleitung durch Schwefelmetalle" (On current conduction in metal sulphides), *Annalen der Physik und Chemie*, **153** : 556–563.

[9] Karl Ferdinand Braun. chem.ch.huji.ac.il

[10] "Diode". Encyclobeamia.solarbotics.net.

6.6   Further reading and external links

- Media related to Electromagnetism at Wikimedia Commons
- The Natural History *Pliny the Elder, The Natural History from Perseus Digital Library*
- The Discovery of the Electron *from the American Institute of Physics*
- Electricity (History) Kenyon College Physics Dept
- Enterprise and electrolysis... *from the Royal Society of Chemistry (chemsoc)*
- Famous Electrochemists *from Eugenii Katz*
    - André-Marie Ampère
    - Gustav Robert Kirchhoff
- SparkMuseum
    - The Development of the Electric Motor
    - Early Telegraph Apparatus
    - Two Kinds of Electrical Fluid: Vitreous and Resinous – 1733 (Charles DuFay),
- Worldwide School (Pure Science-History)
    - Dufay Discovers Vitreous and Resinous Electricity
    - Electricity and Magnetism
    - Faraday and Electro-Magnetic Induction
    - Franklin (Invents the Lightning-Rod, Proves that Lightning Is Electricity, Theory of Electricity)
    - The Leyden Jar Discovered
    - Ludolff's Experiment with the Electric Spark
    - The Origin of the Telegraph
    - Progress in Electricity from Gilbert and Von Guericke to Franklin
    - Storage Batteries
    - William Gilbert and the Study of Magnetism

Albert Einstein in the patent office, Bern Switzerland, 1905

Chapter 7

Timeline of atomic and subatomic physics

A timeline of atomic and subatomic physics.

7.1 Early beginnings

- 600 BCE[1] Kanada theorizes the existence of four kinds of atoms, which could combine to produce diatomic and triatomic molecules.

- 430 BCE[1] Democritus speculates about fundamental indivisible particles—calls them "atoms"

- 200 BCE[1] Jainism calls atom Paramanu which can neither be created nor destroyed. It is eternal, i.e., it existed in the past, exists in the present and will continue to exist in the future. It is the permanent basis of the physical existence. The entire physical existence is composed of these ultimate atoms.

7.2 The beginning of chemistry

- 1766 Henry Cavendish discovers and studies hydrogen

- 1778 Carl Scheele and Antoine Lavoisier discover that air is composed mostly of nitrogen and oxygen

- 1781 Joseph Priestley creates water by igniting hydrogen and oxygen

- 1800 William Nicholson and Anthony Carlisle use electrolysis to separate water into hydrogen and oxygen

- 1803 John Dalton introduces atomic ideas into chemistry and states that matter is composed of atoms of different weights

- 1805 (approximate time) Thomas Young conducts the double-slit experiment with light

- 1811 Amedeo Avogadro claims that equal volumes of gases should contain equal numbers of molecules

- 1832 Michael Faraday states his laws of electrolysis

- 1871 Dmitri Mendeleyev systematically examines the periodic table and predicts the existence of gallium, scandium, and germanium

- 1873 Johannes van der Waals introduces the idea of weak attractive forces between molecules

- 1885 Johann Balmer finds a mathematical expression for observed hydrogen line wavelengths

- 1887 Heinrich Hertz discovers the photoelectric effect

- 1894 Lord Rayleigh and William Ramsay discover argon by spectroscopically analyzing the gas left over after nitrogen and oxygen are removed from air

- 1895 William Ramsay discovers terrestrial helium by spectroscopically analyzing gas produced by decaying uranium

- 1896 Antoine Becquerel discovers the radioactivity of uranium

- 1896 Pieter Zeeman studies the splitting of sodium D lines when sodium is held in a flame between strong magnetic poles

- 1897 J.J. Thomson discovers the electron

- 1898 William Ramsay and Morris Travers discover neon, and negatively charged beta particles

7.3 Timeline of classical mechanics

Main article: Timeline of classical mechanics

7.4 The age of quantum mechanics

- 1887 Heinrich Rudolf Hertz discovers the photoelectric effect that will play a very important role in the development of the quantum theory with Einstein's explanation of this effect in terms of *quanta* of light

- 1896 Wilhelm Conrad Röntgen discovers the X-rays while studying electrons in plasma; scattering X-rays—that were considered as 'waves' of high-energy electromagnetic radiation—Arthur Compton will be able to demonstrate in 1922 the 'particle' aspect of electromagnetic radiation.

- 1900 Paul Villard discovers gamma-rays while studying uranium decay

- 1900 Johannes Rydberg refines the expression for observed hydrogen line wavelengths

- 1900 Max Planck states his quantum hypothesis and blackbody radiation law

- 1902 Philipp Lenard observes that maximum photoelectron energies are independent of illuminating intensity but depend on frequency

- 1902 Theodor Svedberg suggests that fluctuations in molecular bombardment cause the Brownian motion

- 1905 Albert Einstein explains the photoelectric effect

- 1906 Charles Barkla discovers that each element has a characteristic X-ray and that the degree of penetration of these X-rays is related to the atomic weight of the element

- 1909 Hans Geiger and Ernest Marsden discover large angle deflections of alpha particles by thin metal foils

- 1909 Ernest Rutherford and Thomas Royds demonstrate that alpha particles are doubly ionized helium atoms

- 1911 Ernest Rutherford explains the Geiger–Marsden experiment by invoking a nuclear atom model and derives the Rutherford cross section

- 1911 Jean Perrin proves the existence of atoms and molecules

- 1911 Ștefan Procopiu measures the magnetic dipole moment of the electron

- 1912 Max von Laue suggests using crystal lattices to diffract X-rays

- 1912 Walter Friedrich and Paul Knipping diffract X-rays in zinc blende

- 1913 William Henry Bragg and William Lawrence Bragg work out the Bragg condition for strong X-ray reflection

- 1913 Henry Moseley shows that nuclear charge is the real basis for numbering the elements

- 1913 Niels Bohr presents his quantum model of the atom[2]

- 1913 Robert Millikan measures the fundamental unit of electric charge

- 1913 Johannes Stark demonstrates that strong electric fields will split the Balmer spectral line series of hydrogen

- 1914 James Franck and Gustav Hertz observe atomic excitation

- 1914 Ernest Rutherford suggests that the positively charged atomic nucleus contains protons

- 1915 Arnold Sommerfeld develops a modified Bohr atomic model with elliptic orbits to explain relativistic fine structure

- 1916 Gilbert N. Lewis and Irving Langmuir formulate an electron shell model of chemical bonding

- 1917 Albert Einstein introduces the idea of stimulated radiation emission

- 1918 Ernest Rutherford notices that, when alpha particles were shot into nitrogen gas, his scintillation detectors showed the signatures of hydrogen nuclei.

- 1921 Alfred Landé introduces the Landé g-factor

- 1922 Arthur Compton studies X-ray photon scattering by electrons demonstrating the 'particle' aspect of electromagnetic radiation.

- 1922 Otto Stern and Walther Gerlach show "spin quantization"

- 1923 Lise Meitner discovers what is now referred to as the Auger process

- 1924 Louis de Broglie suggests that electrons may have wavelike properties in addition to their 'particle' properties; the *wave–particle duality* has been later extended to all fermions and bosons.

- 1924 John Lennard-Jones proposes a semiempirical interatomic force law

- 1924 Satyendra Bose and Albert Einstein introduce Bose–Einstein statistics

- 1925 Wolfgang Pauli states the quantum exclusion principle for electrons

- 1925 George Uhlenbeck and Samuel Goudsmit postulate electron spin

- 1925 Pierre Auger discovers the Auger process (2 years after Lise Meitner)

- 1925 Werner Heisenberg, Max Born, and Pascual Jordan formulate quantum matrix mechanics

- 1926 Erwin Schrödinger states his nonrelativistic quantum wave equation and formulates quantum wave mechanics

- 1926 Erwin Schrödinger proves that the wave and matrix formulations of quantum theory are mathematically equivalent

- 1926 Oskar Klein and Walter Gordon state their relativistic quantum wave equation, now the Klein–Gordon equation

- 1926 Enrico Fermi discovers the spin–statistics connection, for particles that are now called 'fermions', such as the electron (of spin-1/2).

- 1926 Paul Dirac introduces Fermi–Dirac statistics

- 1926 Gilbert N. Lewis introduces the term "*photon*", thought by him to be "*the carrier of radiant energy.*" [3][4]

- 1927 Clinton Davisson, Lester Germer, and George Paget Thomson confirm the wavelike nature of electrons[5]

- 1927 Werner Heisenberg states the quantum uncertainty principle

- 1927 Max Born interprets the probabilistic nature of wavefunctions

- 1927 Walter Heitler and Fritz London introduce the concepts of valence bond theory and apply it to the hydrogen molecule.

- 1927 Thomas and Fermi develop the Thomas–Fermi model

- 1927 Max Born and Robert Oppenheimer introduce the Born–Oppenheimer approximation

- 1928 Chandrasekhara Raman studies optical photon scattering by electrons

- 1928 Paul Dirac states his relativistic electron quantum wave equation

- 1928 Charles G. Darwin and Walter Gordon solve the Dirac equation for a Coulomb potential

- 1928 Friedrich Hund and Robert S. Mulliken introduce the concept of molecular orbital

- 1929 Oskar Klein discovers the Klein paradox

- 1929 Oskar Klein and Yoshio Nishina derive the Klein–Nishina cross section for high energy photon scattering by electrons

- 1929 Nevill Mott derives the Mott cross section for the Coulomb scattering of relativistic electrons

- 1930 Paul Dirac introduces electron hole theory

- 1930 Erwin Schrödinger predicts the zitterbewegung motion

- 1930 Fritz London explains van der Waals forces as due to the interacting fluctuating dipole moments between molecules

- 1931 John Lennard-Jones proposes the Lennard-Jones interatomic potential

- 1931 Irène Joliot-Curie and Frédéric Joliot observe but misinterpret neutron scattering in paraffin

- 1931 Wolfgang Pauli puts forth the neutrino hypothesis to explain the apparent violation of energy conservation in beta decay

- 1931 Linus Pauling discovers resonance bonding and uses it to explain the high stability of symmetric planar molecules

- 1931 Paul Dirac shows that charge quantization can be explained if magnetic monopoles exist

- 1931 Harold Urey discovers deuterium using evaporation concentration techniques and spectroscopy

- 1932 John Cockcroft and Ernest Walton split lithium and boron nuclei using proton bombardment

- 1932 James Chadwick discovers the neutron

- 1932 Werner Heisenberg presents the proton–neutron model of the nucleus and uses it to explain isotopes

- 1932 Carl D. Anderson discovers the positron

- 1933 Ernst Stueckelberg (1932), Lev Landau (1932), and Clarence Zener discover the Landau–Zener transition

- 1933 Max Delbrück suggests that quantum effects will cause photons to be scattered by an external electric field

- 1934 Irène Joliot-Curie and Frédéric Joliot bombard aluminium atoms with alpha particles to create artificially radioactive phosphorus-30

- 1934 Leó Szilárd realizes that nuclear chain reactions may be possible

- 1934 Enrico Fermi publishes a very successful model of beta decay in which neutrinos were produced.

- 1934 Lev Landau tells Edward Teller that non-linear molecules may have vibrational modes which remove the degeneracy of an orbitally degenerate state (Jahn–Teller effect)

- 1934 Enrico Fermi suggests bombarding uranium atoms with neutrons to make a 93 proton element

- 1934 Pavel Cherenkov reports that light is emitted by relativistic particles traveling in a nonscintillating liquid

- 1935 Hideki Yukawa presents a theory of the nuclear force and predicts the scalar meson

- 1935 Albert Einstein, Boris Podolsky, and Nathan Rosen put forth the EPR paradox

- 1935 Henry Eyring develops the transition state theory

- 1935 Niels Bohr presents his analysis of the EPR paradox

- 1936 Alexandru Proca formulates the relativistic quantum field equations for a massive vector meson of spin-1 as a basis for nuclear forces

- 1936 Eugene Wigner develops the theory of neutron absorption by atomic nuclei

- 1936 Hermann Arthur Jahn and Edward Teller present their systematic study of the symmetry types for which the Jahn–Teller effect is expected[6]

- 1937 Carl Anderson proves experimentally the existence of the pion predicted by Yukawa's theory.

- 1937 Hans Hellmann finds the Hellmann–Feynman theorem

- 1937 Seth Neddermeyer, Carl Anderson, J.C. Street, and E.C. Stevenson discover muons using cloud chamber measurements of cosmic rays

- 1939 Richard Feynman finds the Hellmann–Feynman theorem

- 1939 Otto Hahn and Fritz Strassmann bombard uranium salts with thermal neutrons and discover barium among the reaction products

- 1939 Lise Meitner and Otto Robert Frisch determine that nuclear fission is taking place in the Hahn–Strassmann experiments

- 1942 Enrico Fermi makes the first controlled nuclear chain reaction

- 1942 Ernst Stueckelberg introduces the propagator to positron theory and interprets positrons as negative energy electrons moving backwards through spacetime

- 1943 Sin-Itiro Tomonaga publishes his paper on the basic physical principles of quantum electrodynamics

- 1947 Willis Lamb and Robert Retherford measure the Lamb–Retherford shift

- 1947 Cecil Powell, César Lattes, and Giuseppe Occhialini discover the pi meson by studying cosmic ray tracks

- 1947 Richard Feynman presents his propagator approach to quantum electrodynamics[7]

- 1948 Hendrik Casimir predicts a rudimentary attractive Casimir force on a parallel plate capacitor

- 1951 Martin Deutsch discovers positronium

- 1952 David Bohm propose his interpretation of quantum mechanics

- 1953 Robert Wilson observes Delbruck scattering of 1.33 MeV gamma-rays by the electric fields of lead nuclei

- 1953 Charles H. Townes, collaborating with J. P. Gordon, and H. J. Zeiger, builds the first ammonia maser

- 1954 Chen Ning Yang and Robert Mills investigate a theory of hadronic isospin by demanding local gauge invariance under isotopic spin space rotations, the first non-Abelian gauge theory

- 1955 Owen Chamberlain, Emilio Segrè, Clyde Wiegand, and Thomas Ypsilantis discover the antiproton

- 1956 Frederick Reines and Clyde Cowan detect antineutrino

- 1956 Chen Ning Yang and Tsung Lee propose parity violation by the weak nuclear force

- 1956 Chien Shiung Wu discovers parity violation by the weak force in decaying cobalt

- 1957 Gerhart Luders proves the CPT theorem

- 1957 Richard Feynman, Murray Gell-Mann, Robert Marshak, and E.C.G. Sudarshan propose a vector/axial vector (VA) Lagrangian for weak interactions.[8][9][10][11][12][13]

- 1958 Marcus Sparnaay experimentally confirms the Casimir effect

- 1959 Yakir Aharonov and David Bohm predict the Aharonov–Bohm effect

- 1960 R.G. Chambers experimentally confirms the Aharonov–Bohm effect[14]

- 1961 Murray Gell-Mann and Yuval Ne'eman discover the Eightfold Way patterns, the SU(3) group

- 1961 Jeffrey Goldstone considers the breaking of global phase symmetry

- 1962 Leon Lederman shows that the electron neutrino is distinct from the muon neutrino

- 1963 Eugene Wigner discovers the fundamental roles played by quantum symmetries in atoms and molecules

7.5 The formation and successes of the Standard Model

- 1964 Murray Gell-Mann and George Zweig propose the quark/aces model[15][16]

- 1964 Peter Higgs considers the breaking of local phase symmetry

- 1964 John Stewart Bell shows that all local hidden variable theories must satisfy Bell's inequality

- 1964 Val Fitch and James Cronin observe CP violation by the weak force in the decay of K mesons

- 1967 Steven Weinberg puts forth his electroweak model of leptons[17][18]

- 1969 John Clauser, Michael Horne, Abner Shimony and Richard Holt propose a polarization correlation test of Bell's inequality

- 1970 Sheldon Glashow, John Iliopoulos, and Luciano Maiani propose the charm quark

- 1971 Gerard 't Hooft shows that the Glashow-Salam-Weinberg electroweak model can be renormalized[19]

- 1972 Stuart Freedman and John Clauser perform the first polarization correlation test of Bell's inequality

- 1973 David Politzer and Frank Anthony Wilczek propose the asymptotic freedom of quarks[16]

- 1974 Burton Richter and Samuel Ting discover the J/ψ particle implying the existence of the charm quark

- 1974 Robert J. Buenker and Sigrid D. Peyerimhoff introduce the multireference configuration interaction method.

- 1975 Martin Perl discovers the tau lepton

- 1977 Steve Herb finds the upsilon resonance implying the existence of the beauty/bottom quark

- 1982 Alain Aspect, J. Dalibard, and G. Roger perform a polarization correlation test of Bell's inequality that rules out conspiratorial polarizer communication

- 1983 Carlo Rubbia, Simon van der Meer, and the CERN UA-1 collaboration find the W and Z intermediate vector bosons[20]

- 1989 The Z intermediate vector boson resonance width indicates three quark-lepton generations

- 1994 The CERN LEAR Crystal Barrel Experiment justifies the existence of glueballs (exotic meson).

- 1995 after 18 years searching at Fermilab was discovered the top quark, it had very big mass

- 1998 Super-Kamiokande (Japan) observes evidence for neutrino oscillations, implying that at least one neutrino has mass.

- 1999 Ahmed Zewail wins the Nobel prize in chemistry for his work on femtochemistry for atoms and molecules.[21]

- 2001 The Sudbury Neutrino Observatory (Canada) confirms the existence of neutrino oscillations.

- 2005 At the RHIC accelerator of Brookhaven National Laboratory they have created a quark–gluon liquid of very low viscosity, perhaps the quark–gluon plasma

- 2008 The Large Hadron Collider at CERN is scheduled to begin operation in this year. Its primary goal is to search for the Higgs boson, which has not yet been found.

- 2012 CERN announces the discovery of a new particle with properties consistent with the Higgs boson of the Standard Model after experiments at the Large Hadron Collider.

7.6   Quantum field theories beyond the Standard Model

- 2000 Steven Weinberg. Supersymmetry and Quantum Gravity.[18][22]

- 2003 Leonid Vainerman. Quantum groups, Hopf algebras and quantum field applications.[23]

- Noncommutative quantum field theory

- M.R. Douglas and N. A. Nekrasov (2001) "Noncommutative field theory," Rev. Mod. Phys. 73: 977–1029.

- Szabo, R. J. (2003) "Quantum Field Theory on Noncommutative Spaces," *Physics Reports* 378: 207–99. An expository article on noncommutative quantum field theories.

- Noncommutative quantum field theory, see statistics on arxiv.org

- Seiberg, N. and E. Witten (1999) "String Theory and Noncommutative Geometry," *Journal of High Energy Physics*

- Sergio Doplicher, Klaus Fredenhagen and John Roberts, Sergio Doplicher, Klaus Fredenhagen, John E. Roberts (1995) The quantum structure of spacetime at the Planck scale and quantum fields," *Commun. Math. Phys.* 172: 187–220.

- Alain Connes (1994) *Noncommutative geometry.* Academic Press. ISBN 0-12-185860-X.

- -------- (1995) "Noncommutative geometry and reality", *J. Math. Phys.* 36: 6194.

- -------- (1996) "Gravity coupled with matter and the foundation of noncommutative geometry," *Comm. Math. Phys.* 155: 109.

- -------- (2006) "Noncommutative geometry and physics,"

- -------- and M. Marcolli, *Noncommutative Geometry: Quantum Fields and Motives.* American Mathematical Society (2007).

- Chamseddine, A., A. Connes (1996) "The spectral action principle," *Comm. Math. Phys.* 182: 155.

- Chamseddine, A., A. Connes, M. Marcolli (2007) "Gravity and the Standard Model with neutrino mixing," *Adv. Theor. Math. Phys.* 11: 991.

- Jureit, Jan-H., Thomas Krajewski, Thomas Schücker, and Christoph A. Stephan (2007) "On the noncommutative standard model," *Acta Phys. Polon.* B38: 3181–3202.

- Schücker, Thomas (2005) *Forces from Connes's geometry.* Lecture Notes in Physics 659, Springer.

- Noncommutative standard model

- Noncommutative geometry

7.7   See also

- History of subatomic physics

- History of quantum mechanics

- History of quantum field theory

- History of the molecule

- History of thermodynamics

- History of chemistry

- Golden age of physics

7.8   References

[1] Teresi, Dick (2010). *Lost Discoveries: The Ancient Roots of Modern Science.* Simon and Schuster. pp. 213–214. ISBN 978-1-4391-2860-2.

[2] Jammer, Max (1966), *The conceptual development of quantum mechanics*, New York: McGraw-Hill, OCLC 534562

[3] Gilbert N. Lewis. Letter to the editor of *Nature* (Vol. 118, Part 2, December 18, 1926, pp. 874–875).

[4] The origin of the word "photon"

[5] The Davisson–Germer experiment, which demonstrates the wave nature of the electron

[6] A. Abragam and B. Bleaney. 1970. Electron Parmagnetic Resonance of Transition Ions, Oxford University Press: Oxford, U.K., p. 911

[7] Feynman, R.P. (2006) [1985]. *QED: The Strange Theory of Light and Matter.* Princeton University Press. ISBN 0-691-12575-9.

[8] Richard Feynman; **QED**. Princeton University Press: Princeton, (1982)

[9] Richard Feynman; *Lecture Notes in Physics.* Princeton University Press: Princeton, (1986)

[10] Feynman, R.P. (2001) [1964]. *The Character of Physical Law.* MIT Press. ISBN 0-262-56003-8.

[11] Feynman, R.P. (2006) [1985]. *QED: The Strange Theory of Light and Matter.* Princeton University Press. ISBN 0-691-12575-9.

[12] Schweber, Silvan S. ; Q.E.D. and the men who made it: Dyson, Feynman, Schwinger, and Tomonaga, Princeton University Press (1994) [ISBN 0-691-03327-7]

[13] Schwinger, Julian ; Selected Papers on Quantum Electrodynamics, Dover Publications, Inc. (1958) [ISBN 0-486-60444-6]

[14]   • Kleinert, H. (2008). *Multivalued Fields in Condensed Matter, Electrodynamics, and Gravitation* (PDF). World Scientific. ISBN 978-981-279-170-2.

[15] Yndurain, Francisco Jose ; *Quantum Chromodynamics: An Introduction to the Theory of Quarks and Gluons*, Springer Verlag, New York, 1983. [ISBN 0-387-11752-0]

[16] Frank Wilczek (1999) "Quantum field theory", *Reviews of Modern Physics* 71: S83–S95. Also doi=10.1103/Rev. Mod. Phys. 71.

[17] Weinberg, Steven ; The Quantum Theory of Fields: Foundations (vol. I), Cambridge University Press (1995) [ISBN 0-521-55001-7] The first chapter (pp. 1–40) of Weinberg's monumental treatise gives a brief history of Q.F.T., pp. 608.

[18] Weinberg, Steven; The Quantum Theory of Fields: Modern Applications (vol. II), Cambridge University Press:Cambridge, U.K. (1996) [ISBN 0-521-55001-7], pp. 489.

[19]   • Gerard 't Hooft (2007) "The Conceptual Basis of Quantum Field Theory" in Butterfield, J., and John Earman, eds., *Philosophy of Physics, Part A*. Elsevier: 661-730.

[20] Pais, Abraham ; Inward Bound: Of Matter & Forces in the Physical World, Oxford University Press (1986) [ISBN 0-19-851997-4] Written by a former Einstein assistant at Princeton, this is a beautiful detailed history of modern fundamental physics, from 1895 (discovery of X-rays) to 1983 (discovery of vectors bosons at C.E.R.N.)

[21] "Press Release: The 1999 Nobel Prize in Chemistry". 12 October 1999. Retrieved 30 June 2013.

[22] Weinberg, Steven; The Quantum Theory of Fields: Supersymmetry (vol. III), Cambridge University Press:Cambridge, U.K. (2000) [ISBN 0-521-55002-5], pp. 419.

[23] Leonid Vainerman, editor. 2003. *Locally Compact Quantum Groups and Groupoids. Proceed. Theor. Phys. Strassbourg in 2002*, Walter de Gruyter: Berlin and New York

7.9 External links

- Alain Connes official website with downloadable papers.
- Alain Connes's Standard Model.
- A History of Quantum Mechanics
- A Brief History of Quantum Mechanics

Chapter 8

Timeline of particle physics

The **timeline of particle physics** lists the sequence of particle physics theories and discoveries in chronological order. The most modern developments follow the scientific development of the discipline of particle physics.

8.1   19th century

- 1815 – William Prout hypothesizes that all matter is built up from hydrogen, adumbrating the proton;

- 1838 – Richard Laming hypothesized a subatomic particle carrying electric charge;

- 1858 – Julius Plücker produced cathode rays;

- 1874 – George Johnstone Stoney hypothesizes a minimum unit of electric charge. In 1891, he coins the word electron for it;

- 1886 – Eugene Goldstein produced anode rays;

- 1897 – J. J. Thomson discovered the electron;

- 1899 – Ernest Rutherford discovered the alpha and beta particles emitted by uranium;

- 1900 – Paul Villard discovered the gamma ray in uranium decay.

8.2   20th century

- 1905 – Albert Einstein hypothesized the photon to explain the photoelectric effect.

- 1911 – Ernest Rutherford discovered the nucleus of an atom;

- 1919 – Ernest Rutherford discovered the proton;

- 1928 – Paul Dirac postulated the existence of positrons as a consequence of the Dirac equation;

- 1930 – Wolfgang Pauli postulated the neutrino to explain the energy spectrum of beta decays;

- 1932 – James Chadwick discovered the neutron;

- 1932 – Carl D. Anderson discovered the positron;

- 1935 – Hideki Yukawa predicted the existence of mesons as the carrier particles of the strong nuclear force;

- 1936 – Carl D. Anderson discovered the muon while he studied cosmic radiation;

- 1947 – George Dixon Rochester and Clifford Charles Butler discovered the kaon, the first strange particle;

- 1947 – Cecil Powell, César Lattes and Giuseppe Occhialini discovered the pion;

- 1955 – Owen Chamberlain, Emilio Segrè, Clyde Wiegand, and Thomas Ypsilantis discovered the antiproton;

- 1956 – Clyde Cowan and Frederick Reines discovered the (electron) neutrino;

- 1957 – Bruno Pontecorvo postulated the flavor oscillation;

- 1962 – Leon M. Lederman, Melvin Schwartz and Jack Steinberger discovered the muon neutrino;

- 1967 – Bruno Pontecorvo postulated neutrino oscillation;

- 1974 – Burton Richter and Samuel Ting discovered the J/ψ particle composed of charm quarks;

- 1977 – Upsilon particle discovered at Fermilab, demonstrating the existence of the bottom quark;

- 1977 – Martin Lewis Perl discovered the tau lepton after a series of experiments;

- 1979 – Gluon observed indirectly in three-jet events at DESY;

- 1983 – Carlo Rubbia and Simon van der Meer discovered the W and Z bosons;

- 1995 – Top quark discovered at Fermilab;

8.3   21st century

- 2000 – Tau neutrino proved distinct from other neutrinos at Fermilab.

- 2012 – Higgs boson-like particle discovered at CERN's Large Hadron Collider (LHC).

8.4   See also

- Particle physics

- Timeline of particle physics technology

- Timeline of cosmology

- Timeline of the Big Bang

Chapter 9

Timeline of particle discoveries

This is a **timeline of subatomic particle discoveries**, including all particles thus far discovered which appear to be elementary (that is, indivisible) given the best available evidence. It also includes the discovery of composite particles and antiparticles that were of particular historical importance.

More specifically, the inclusion criteria are:

- Elementary particles from the Standard Model of particle physics that have so far been observed. The Standard Model is the most comprehensive existing model of particle behavior. All Standard Model particles including the Higgs boson have been verified, and all other observed particles are combinations of two or more Standard Model particles.

- Antiparticles which were historically important to the development of particle physics, specifically the positron and antiproton. The discovery of these particles required very different experimental methods from that of their ordinary matter counterparts, and provided evidence that *all* particles had antiparticles—an idea that is fundamental to quantum field theory, the modern mathematical framework for particle physics. In the case of most subsequent particle discoveries, the particle and its anti-particle were discovered essentially simultaneously.

- Composite particles which were the first particle discovered containing a particular elementary constituent, or whose discovery was critical to the understanding of particle physics.

Time	Event
1800	William Herschel discovers "heat rays"
1801	Johann Wilhelm Ritter made the hallmark observation that invisible rays just beyond the violet end of the visible spectrum were especially effective at lightening silver chloride-soaked paper. He called them "oxidizing rays" to emphasize chemical reactivity and to distinguish them from "heat rays" at the other end of the invisible spectrum (both of which were later determined to be photons). The more general term "chemical rays" was adopted shortly thereafter to describe the oxidizing rays, and it remained popular throughout the 19th century. The terms chemical and heat rays were eventually dropped in favor of **ultraviolet** and **infrared** radiation, respectively.[1]
1895	Discovery of the ultraviolet radiation below 200 nm, named **vacuum ultraviolet** (later identified as photons) because it is strongly absorbed by air, by the German physicist Victor Schumann[2]
1895	X-ray produced by Wilhelm Röntgen (later identified as photons)[3]
1897	**Electron** discovered by J. J. Thomson[4]
1899	**Alpha particle** discovered by Ernest Rutherford in uranium radiation[5]
1900	**Gamma ray** (a high-energy photon) discovered by Paul Villard in uranium decay[6]
1911	**Atomic nucleus** identified by Ernest Rutherford, based on scattering observed by Hans Geiger and Ernest Marsden[7]
1919	**Proton** discovered by Ernest Rutherford[8]
1932	**Neutron** discovered by James Chadwick[9] (predicted by Rutherford in 1920[10])
1932	**Antielectron** (or **positron**), the first antiparticle, discovered by Carl D. Anderson[11] (proposed by Paul Dirac in 1927 and by Ettore Majorana in 1928)
1937	**Muon** (or **mu lepton**) discovered by Seth Neddermeyer, Carl D. Anderson, J.C. Street, and E.C. Stevenson, using cloud chamber measurements of cosmic rays[12] (it was mistaken for the pion until 1947[13])
1947	**Pion** (or **pi meson**) discovered by C. F. Powell's group (predicted by Hideki Yukawa in 1935[14])
1947	**Kaon** (or **K meson**), the first strange particle, discovered by George Dixon Rochester and Clifford Charles Butler[15]
1947	Λ^0 discovered during a study of cosmic-ray interactions[16]
1955	**Antiproton** discovered by Owen Chamberlain, Emilio Segrè, Clyde Wiegand, and Thomas Ypsilantis[17]

1956	**Electron antineutrino** detected by Frederick Reines and Clyde Cowan (proposed by Wolfgang Pauli in 1930 to explain the apparent violation of energy conservation in beta decay)[18] At the time it was simply referred to as *neutrino* since there was only one known neutrino.
1962	**Muon neutrino** (or **mu neutrino**) shown to be distinct from the electron neutrino by a group headed by Leon Lederman[19]
1964	**Xi baryon** discovery at Brookhaven National Laboratory[20]
1969	**Partons** (internal constituents of hadrons) observed in deep inelastic scattering experiments between protons and electrons at SLAC;[21][22] this was eventually associated with the quark model (predicted by Murray Gell-Mann and George Zweig in 1964) and thus constitutes the discovery of the **up quark**, **down quark**, and **strange quark**.
1974	**J/ψ meson** discovered by groups headed by Burton Richter and Samuel Ting, demonstrating the existence of the **charm quark**[23][24] (proposed by James Bjorken and Sheldon Lee Glashow in 1964[25])
1975	**Tau** discovered by a group headed by Martin Perl[26]
1977	**Upsilon meson** discovered at Fermilab, demonstrating the existence of the **bottom quark**[27] (proposed by Kobayashi and Maskawa in 1973)
1979	**Gluon** observed indirectly in three-jet events at DESY[28]
1983	**W and Z bosons** discovered by Carlo Rubbia, Simon van der Meer, and the CERN UA1 collaboration[29][30] (predicted in detail by Sheldon Glashow, Mohammad Abdus Salam, and Steven Weinberg)
1995	**Top quark** discovered at Fermilab[31][32]
1995	**Antihydrogen** produced and measured by the LEAR experiment at CERN[33]
2000	**Tau neutrino** first observed directly at Fermilab[34]
2011	**Antihelium-4** produced and measured by the STAR detector; the first particle to be discovered by the experiment
2012	A particle exhibiting most of the predicted characteristics of the **Higgs boson** discovered by researchers conducting the Compact Muon Solenoid and ATLAS experiments at CERN's Large Hadron Collider[35]

[14] C.D. Anderson (1935). "On the Interaction of Elementary Particles". *Proceedings of the Physico-Mathematical Society of Japan* **17**: 48.

[15] G.D. Rochester, C.C. Butler (1947). "Evidence for the Existence of New Unstable Elementary Particles". *Nature* **160** (4077): 855–857. Bibcode:1947Natur.160..855R. doi:10.1038/160855a0.

[16] The Strange Quark

[17] O. Chamberlain, E. Segrè, C. Wiegand, T. Ypsilantis (1955). "Observation of Antiprotons". *Physical Review* **100** (3): 947–950. Bibcode:1955PhRv..100..947C. doi:10.1103/PhysRev.100.947.

[18] F. Reines, C.L. Cowan (1956). "The Neutrino". *Nature* **178**(4531): 446–449.Bibcode:1956Natur.178..446R.doi:10.1038/17

[19] G. Danby; et al. (1962). "Observation of High-Energy Neutrino Reactions and the Existence of Two Kinds of Neutrinos". *Physical Review Letters* **9** (1): 36–44. Bibcode:1962PhRvL...9...36D. doi:10.1103/PhysRevLett.9.36.

[20] R. Nave. "The Xi Baryon". Hyperphysics. Retrieved 20 June 2009.

[21] E.D. Bloom; et al. (1969). "High-Energy Inelastic e–p Scattering at 6° and 10°". *Physical Review Letters* **23** (16): 930–934. Bibcode:1969PhRvL..23..930B. doi:10.1103/PhysRevLett.23.930.

[22] M. Breidenbach; et al. (1969). "Observed Behavior of Highly Inelastic Electron-Proton Scattering". *Physical Review Letters* **23** (16): 935–939. Bibcode:1969PhRvL..23..935B. doi:10.1103/PhysRevLett.23.935.

[23] J.J. Aubert; et al. (1974). "Experimental Observation of a Heavy Particle *J*". *Physical Review Letters* **33** (23): 1404–1406. Bibcode:1974PhRvL..33.1404A. doi:10.1103/PhysRevLett.33.1404.

[24] J.-E. Augustin; et al. (1974). "Discovery of a Narrow Resonance in e^+e^- Annihilation". *Physical Review Letters* **33** (23): 1406–1408. Bibcode:1974PhRvL..33.1406A. doi:10.1103/PhysRevLett.33.1406.

[25] B.J. Bjørken, S.L. Glashow (1964). "Elementary Particles and SU(4)".*Physics Letters***11**(3): 255–257.Bibcode:1964PhL.. doi:10.1016/0031-9163(64)90433-0.

[26] M.L. Perl; et al. (1975). "Evidence for Anomalous Lepton Production in e^+–e^- Annihilation". *Physical Review Letters* **35** (22): 1489–1492. Bibcode:1975PhRvL..35.1489P. doi:10.1103/PhysRevLett.35.1489.

[27] S.W. Herb; et al. (1977). "Observation of a Dimuon Resonance at 9.5 GeV in 400-GeV Proton-Nucleus Collisions". *Physical Review Letters* **39** (5): 252–255. Bibcode:1977PhRvL..39..252H. doi:10.1103/PhysRevLett.39.252.

[28] D.P. Barber; et al. (1979). "Discovery of Three-Jet Events and a Test of Quantum Chromodynamics at PETRA". *Physical Review Letters* **43** (12): 830–833. Bibcode:1979PhRvL..43..830B. doi:10.1103/PhysRevLett.43.830.

[29] J.J. Aubert *et al.* (European Muon Collaboration) (1983). "The ratio of the nucleon structure functions F_2^N for iron and deuterium". *Physics Letters B* **123** (3–4): 275–278. Bibcode:1983PhLB..123..275A. doi:10.1016/0370-2693(83)90437-9.

[30] G. Arnison *et al.* (UA1 collaboration) (1983). "Experimental observation of lepton pairs of invariant mass around 95 GeV/c^2 at the CERN SPS collider". *Physics Letters B* **126** (5): 398–410. Bibcode:1983PhLB..126..398A. doi:10.1016/0370-2693(83)90 188-0.

[31] F. Abe *et al.* (CDF collaboration) (1995). "Observation of Top quark production in p–p Collisions with the Collider Detector at Fermilab". *Physical Review Letters* **74** (14): 2626–2631. arXiv:hep-ex/9503002. Bibcode:1995PhRvL..74.2626A. doi:10.1103/PhysRevLett.74.2626. PMID 10057978.

[32] S. Arabuchi *et al.* (D0 collaboration) (1995). "Observation of the Top Quark". *Physical Review Letters* **74** (14): 2632–2637. arXiv:hep-ex/9503003. Bibcode:1995PhRvL..74.2632A. doi:10.1103/PhysRevLett.74.2632. PMID 10057979.

[33] G. Baur; et al. (1996). "Production of Antihydrogen". *Physics Letters B* **368** (3): 251–258. Bibcode:1996PhLB..368..251B. doi:10.1016/0370-2693(96)00005-6.

[34] "Physicists Find First Direct Evidence for Tau Neutrino at Fermilab" (Press release). Fermilab. 20 July 2000. Retrieved 20 March 2010.

[35] Boyle, Alan (4 July 2012). "Milestone in Higgs quest: Scientists find new particle". *MSNBC* (MSNBC). Retrieved 5 July 2012.

- V.V. Ezhela; et al. (1996). *Particle Physics: One Hundred Years of Discoveries: An Annotated Chronological Bibliography.* Springer–Verlag. ISBN 1-56396-642-5.

Chapter 10

Timeline of states of matter and phase transitions

Timeline of states of matter and phase transitions

- 1895 – Pierre Curie discovers that induced magnetization is proportional to magnetic field strength

- 1911 – Heike Kamerlingh Onnes discloses his research on superconductivity

- 1912 – Peter Debye derives the T-cubed law for the low temperature heat capacity of a nonmetallic solid

- 1925 – Ernst Ising presents the solution to the one-dimensional Ising model

- 1928 – Felix Bloch applies quantum mechanics to electrons in crystal lattices, establishing the quantum theory of solids

- 1929 – Paul Adrien Maurice Dirac and Werner Karl Heisenberg develop the quantum theory of ferromagnetism

- 1932 – Louis Eugène Félix Néel discovers antiferromagnetism

- 1933 – Walther Meissner and Robert Ochsenfeld discover perfect superconducting diamagnetism

- 1933–1937 – Lev Davidovich Landau develops the Landau theory of phase transitions

- 1937 – Pyotr Leonidovich Kapitsa and John Frank Allen discover superfluidity

- 1941 – Lev Davidovich Landau explains superfluidity

- 1942 – Hannes Alfvén predicts magnetohydrodynamic waves in plasmas

- 1944 – Lars Onsager publishes the exact solution to the two-dimensional Ising model

- 1957 – John Bardeen, Leon Cooper, and Robert Schrieffer develop the BCS theory of superconductivity

- *End of the 50s* – Lev Davidovich Landau develops the theory of Fermi liquid

- 1959 – Philip Warren Anderson predicts localization in disordered systems

- 1972 – Douglas Osheroff, Robert C. Richardson, and David Lee discover that helium-3 can become a superfluid

- 1974 – Kenneth G. Wilson develops the renormalization group technique for treating phase transitions

- 1980 – Klaus von Klitzing discovers the quantum Hall effect

- 1982 – Horst L. Stoermer and Daniel C. Tsui discover the fractional quantum Hall effect

- 1983 – Robert B. Laughlin explains the fractional quantum Hall effect
- 1987 – Karl Alexander Müller and Georg Bednorz discover high critical temperature ceramic superconductors

Chapter 11

Timeline of thermodynamics

A timeline of events related to thermodynamics.

11.1 Before 1800

- 1650 – Otto von Guericke builds the first vacuum pump

- 1660 – Robert Boyle experimentally discovers Boyle's Law, relating the pressure and volume of a gas (published 1662)

- 1665 – Robert Hooke stated: "Heat being nothing else but a very brisk and vehement agitation of the parts of a body."

- 1669 – J.J. Becher puts forward a theory of combustion involving *combustible earth* (Latin *terra pinguis*).

- 1676–1689 – Gottfried Leibniz develops the concept of *vis viva*, a limited version of the conservation of energy

- 1679 – Denis Papin designed a steam digester which inspired the development of the piston-and-cylinder steam engine.

- 1694–1734 – Georg Ernst Stahl names Becher's combustible earth as phlogiston and develops the theory

- 1698 – Thomas Savery patents an early steam engine

- 1702 – Guillaume Amontons introduces the concept of absolute zero, based on observations of gases

- 1738 – Daniel Bernoulli publishes *Hydrodynamica*, initiating the kinetic theory

- 1749 – Émilie du Châtelet, in her French translation and commentary on Newton's *Philosophiae Naturalis Principia Mathematica*, derives the conservation of energy from the first principles of Newtonian mechanics.

- 1761 – Joseph Black discovers that ice absorbs heat without changing its temperature when melting

- 1772 – Black's student Daniel Rutherford discovers nitrogen, which he calls *phlogisticated air*, and together they explain the results in terms of the phlogiston theory

- 1776 – John Smeaton publishes a paper on experiments related to power, work, momentum, and kinetic energy, supporting the conservation of energy

- 1777 – Carl Wilhelm Scheele distinguishes heat transfer by thermal radiation from that by convection and conduction

- 1783 – Antoine Lavoisier discovers oxygen and develops an explanation for combustion; in his paper "Réflexions sur le phlogistique", he deprecates the phlogiston theory and proposes a caloric theory

- 1784 – Jan Ingenhousz describes Brownian motion of charcoal particles on water

- 1791 – Pierre Prévost shows that all bodies radiate heat, no matter how hot or cold they are

- 1798 – Count Rumford (Benjamin Thompson) performs measurements of the frictional heat generated in boring cannons and develops the idea that heat is a form of kinetic energy; his measurements are inconsistent with caloric theory, but are also sufficiently imprecise as to leave room for doubt.

11.2 1800–1847

- 1802 – Joseph Louis Gay-Lussac publishes Charles's law, discovered (but unpublished) by Jacques Charles around 1787; this shows the dependency between temperature and volume. Gay-Lussac also formulates the law relating temperature with pressure (the pressure law, or Gay-Lussac's law)

- 1804 – Sir John Leslie observes that a matte black surface radiates heat more effectively than a polished surface, suggesting the importance of black body radiation

- 1805 – William Hyde Wollaston defends the conservation of energy in *On the Force of Percussion*

- 1808 – John Dalton defends caloric theory in *A New System of Chemistry* and describes how it combines with matter, especially gases; he proposes that the heat capacity of gases varies inversely with atomic weight

- 1810 – Sir John Leslie freezes water to ice artificially

- 1813 – Peter Ewart supports the idea of the conservation of energy in his paper *On the measure of moving force*; the paper strongly influences Dalton and his pupil, James Joule

- 1819 – Pierre Louis Dulong and Alexis Thérèse Petit give the Dulong-Petit law for the specific heat capacity of a crystal

- 1820 – John Herapath develops some ideas in the kinetic theory of gases but mistakenly associates temperature with molecular momentum rather than kinetic energy; his work receives little attention other than from Joule

- 1822 – Joseph Fourier formally introduces the use of dimensions for physical quantities in his *Théorie Analytique de la Chaleur*

- 1822 – Marc Seguin writes to John Herschel supporting the conservation of energy and kinetic theory

- 1824 – Sadi Carnot analyzes the efficiency of steam engines using caloric theory; he develops the notion of a reversible process and, in postulating that no such thing exists in nature, lays the foundation for the second law of thermodynamics, and initiating the science of thermodynamics

- 1827 – Robert Brown discovers the Brownian motion of pollen and dye particles in water

- 1831 – Macedonio Melloni demonstrates that black body radiation can be reflected, refracted, and polarised in the same way as light

- 1834 – Émile Clapeyron popularises Carnot's work through a graphical and analytic formulation. He also combined Boyle's Law, Charles's Law, and Gay-Lussac's Law to produce a Combined Gas Law. PV/T = k

- 1841 – Julius Robert von Mayer, an amateur scientist, writes a paper on the conservation of energy, but his lack of academic training leads to its rejection

- 1842 – Mayer makes a connection between work, heat, and the human metabolism based on his observations of blood made while a ship's surgeon; he calculates the mechanical equivalent of heat

- 1842 – William Robert Grove demonstrates the thermal dissociation of molecules into their constituent atoms, by showing that steam can be disassociated into oxygen and hydrogen, and the process reversed

- 1843 – John James Waterston fully expounds the kinetic theory of gases, but is ridiculed and ignored

- 1843 – James Joule experimentally finds the mechanical equivalent of heat

- 1845 – Henri Victor Regnault added Avogadro's Law to the Combined Gas Law to produce the Ideal Gas Law. PV = nRT

- 1846 – Karl-Hermann Knoblauch publishes *De calore radiante disquisitiones experimentis quibusdam novis illustratae*

- 1846 – Grove publishes an account of the general theory of the conservation of energy in *On The Correlation of Physical Forces*

- 1847 – Hermann von Helmholtz publishes a definitive statement of the conservation of energy, the first law of thermodynamics

11.3   1848–1899

- 1848 – William Thomson extends the concept of absolute zero from gases to all substances

- 1849 – William John Macquorn Rankine calculates the correct relationship between saturated vapour pressure and temperature using his *hypothesis of molecular vortices*

- 1850 – Rankine uses his *vortex* theory to establish accurate relationships between the temperature, pressure, and density of gases, and expressions for the latent heat of evaporation of a liquid; he accurately predicts the surprising fact that the apparent specific heat of saturated steam will be negative.

- 1850 – Rudolf Clausius gives the first clear joint statement of the first and second law of thermodynamics, abandoning the caloric theory, but preserving Carnot's principle.

- 1851 – Thomson gives an alternative statement of the second law.

- 1852 – Joule and Thomson demonstrate that a rapidly expanding gas cools, later named the Joule–Thomson effect or Joule–Kelvin effect

- 1854 – Helmholtz puts forward the idea of the heat death of the universe

- 1854 – Clausius establishes the importance of dQ/T (Clausius's theorem), but does not yet name the quantity.

- 1854 – Rankine introduces his *thermodynamic function*, later identified as entropy

- 1856 – August Krönig publishes an account of the kinetic theory of gases, probably after reading Waterston's work

- 1857 – Clausius gives a modern and compelling account of the kinetic theory of gases in his *On the nature of motion called heat*

- 1859 – James Clerk Maxwell discovers the distribution law of molecular velocities

- 1859 – Gustav Kirchhoff shows that energy emission from a black body is a function of only temperature and frequency

- 1862 – "Disgregation," a precursor of entropy, was defined in 1862 by Rudolf Clausius as the magnitude of the degree of separation of molecules of a body

- 1865 – Clausius introduces the modern macroscopic concept of entropy

- 1865 – Josef Loschmidt applies Maxwell's theory to estimate the number-density of molecules in gases, given observed gas viscosities.

- 1867 – Maxwell asks whether Maxwell's demon could reverse irreversible processes

- 1870 – Clausius proves the scalar virial theorem

- 1872 – Ludwig Boltzmann states the Boltzmann equation for the temporal development of distribution functions in phase space, and publishes his H-theorem

- 1873 - Van der Waal gave his famous equation of state

- 1874 – Thomson formally states the second law of thermodynamics.

- 1876 – Josiah Willard Gibbs publishes the first of two papers (the second appears in 1878) which discuss phase equilibria, statistical ensembles, the free energy as the driving force behind chemical reactions, and chemical thermodynamics in general.

- 1876 – Loschmidt criticises Boltzmann's H theorem as being incompatible with microscopic reversibility (Loschmidt's paradox).

- 1877 – Boltzmann states the relationship between entropy and probability.

- 1879 – Jožef Stefan observes that the total radiant flux from a blackbody is proportional to the fourth power of its temperature and states the Stefan–Boltzmann law.

- 1884 – Boltzmann derives the Stefan–Boltzmann blackbody radiant flux law from thermodynamic considerations.

- 1888 – Henri-Louis Le Chatelier states his principle that the response of a chemical system perturbed from equilibrium will be to counteract the perturbation.

- 1889 – Walther Nernst relates the voltage of electrochemical cells to their chemical thermodynamics via the Nernst equation.

- 1889 – Svante Arrhenius introduces the idea of activation energy for chemical reactions, giving the Arrhenius equation.

- 1893 – Wilhelm Wien discovers the displacement law for a blackbody's maximum specific intensity.

11.4 1900–1944

- 1900 – Max Planck suggests that light may be emitted in discrete frequencies, giving his law of black-body radiation

- 1905 – Albert Einstein argues that the reality of quanta would explain the photoelectric effect

- 1905 – Einstein mathematically analyzes Brownian motion as a result of random molecular motion

- 1906 – Nernst presents a formulation of the third law of thermodynamics

- 1907 – Einstein uses quantum theory to estimate the heat capacity of an Einstein solid

- 1909 – Constantin Carathéodory develops an axiomatic system of thermodynamics

- 1910 – Einstein and Marian Smoluchowski find the Einstein–Smoluchowski formula for the attenuation coefficient due to density fluctuations in a gas

- 1911 – Paul Ehrenfest and Tatjana Ehrenfest–Afanassjewa publish their classical review on the statistical mechanics of Boltzmann, *Begriffliche Grundlagen der statistischen Auffassung in der Mechanik*

- 1912 – Peter Debye gives an improved heat capacity estimate by allowing low-frequency phonons

- 1916 – Sydney Chapman and David Enskog systematically develop the kinetic theory of gases.

- 1916 – Einstein considers the thermodynamics of atomic spectral lines and predicts stimulated emission

- 1919 – James Jeans discovers that the dynamical constants of motion determine the distribution function for a system of particles

- 1920 – Megh Nad Saha states his ionization equation

- 1923 – Debye and Erich Huckel publish a statistical treatment of the dissociation of electrolytes

- 1924 – Satyendra Nath Bose introduces Bose–Einstein statistics, in a paper translated by Einstein

- 1926 – Enrico Fermi and Paul Dirac introduce Fermi–Dirac statistics for fermions

- 1927 – John von Neumann introduces the density matrix representation and establishes quantum statistical mechanics

- 1928 – John B. Johnson discovers Johnson noise in a resistor

- 1928 – Harry Nyquist derives the fluctuation-dissipation theorem, a relationship to explain Johnson noise in a resistor

- 1929 – Lars Onsager derives the Onsager reciprocal relations

- 1938 – Anatoly Vlasov proposes the Vlasov equation for a correct dynamical description of ensembles of particles with collective long range interaction.

- 1939 – Nikolay Krylov and Nikolay Bogolyubov give the first consistent microscopic derivation of the Fokker-Planck equation in the single scheme of classical and quantum mechanics.

- 1942 – Joseph L. Doob states his theorem on Gauss–Markov processes

- 1944 – Lars Onsager gives an analytic solution to the 2-dimensional Ising model, including its phase transition

11.5   1945–present

- 1945–1946 – Nikolay Bogoliubov develops a general method for a microscopic derivation of kinetic equations for classical statistical systems using BBGKY hierarchy.

- 1947 – Nikolay Bogoliubov and Kirill Gurov extend this method for a microscopic derivation of kinetic equations for quantum statistical systems.

- 1948 – Claude Elwood Shannon establishes information theory.

- 1957 – Aleksandr Solomonovich Kompaneets derives his Compton scattering Fokker–Planck equation.

- 1957 – Ryogo Kubo derives the first of the Green-Kubo relations for linear transport coefficients.

- 1957 – Edwin T. Jaynes gives MaxEnt interpretation of thermodynamics from information theory.

- 1960–1965 – Dmitry Zubarev develops the method of non-equilibrium statistical operator, which becomes a classical tool in the statistical theory of non-equilibrium processes.

- 1972 – Jacob Bekenstein suggests that black holes have an entropy proportional to their surface area.

- 1974 – Stephen Hawking predicts that black holes will radiate particles with a black-body spectrum which can cause black hole evaporation

- 1977 – Ilya Prigogine wins the Nobel prize for his work on dissipative structures in thermodynamic systems far from equilibrium. The importation and dissipation of energy could reverse the 2nd law of thermodynamics.

11.6   See also

- History of physics

- History of thermodynamics

- Timeline of information theory

- List of notable textbooks in statistical mechanics

Chapter 12

Timeline of gravitational physics and relativity

Timeline of gravitational physics and general relativity

12.1   Before 1500

- 3rd century BC - Aristarchus of Samos proposes heliocentric model, measures the distance to the Moon and its size

12.2   1500s

- 1543 – Nicolaus Copernicus places the Sun at the gravitational center, starting a revolution in science

- 1583 – Galileo Galilei induces the period relationship of a pendulum from observation (according to later biographer).

- 1586 - Simon Stevin demonstrates that two objects of different mass accelerate at the same rate when dropped.

- 1589 – Galileo Galilei describes a hydrostatic balance for measuring specific gravity.

- 1590 – Galileo Galilei formulates modified Aristotelean theory of motion (later retracted) based on density rather than weight of objects.

12.3   1600s

- 1602 – Galileo Galilei conducts experiments on pendulum motion.

- 1604 – Galileo Galilei conducts experiments with inclined planes and induces the law of falling objects.

- 1607 – Galileo Galilei arrives a mathematical formulation of the law of falling objects based on his earlier experiments.

- 1608 – Galileo Galilei discovers the parabolic arc of projectiles through experiment.

- 1609 - Johannes Kepler describes to motion of planets around the Sun, now known as Kepler's laws of planetary motion.

- 1640 – Ismaël Bullialdus suggests an inverse-square gravitational force law.

- 1665 – Isaac Newton introduces an inverse-square universal law of gravitation uniting terrestrial and celestial theories of motion and uses it to predict the orbit of the Moon and the parabolic arc of projectiles.

- 1684 – Isaac Newton proves that planets moving under an inverse-square force law will obey Kepler's laws

- 1686 – Isaac Newton uses a fixed length pendulum with weights of varying composition to test the weak equivalence principle to 1 part in 1000

12.4   1700s

- 1798 – Henry Cavendish measures the force of gravity between two masses, leading to the first accurate value for the gravitational constant

12.5   1800s

- 1846 – Urbain Le Verrier and John Couch Adams, studying Uranus orbit, independently prove that another, farther planet must exist. Neptune was found at the predicted moment and position.

- 1855 – Le Verrier observes a 35 arcsecond per century excess precession of Mercury's orbit and attributes it to another planet, inside Mercury's orbit. The planet was never found. See Vulcan.

- 1876 – William Kingdon Clifford suggests that the motion of matter may be due to changes in the geometry of space

- 1882 – Simon Newcomb observes a 43 arcsecond per century excess precession of Mercury's orbit

- 1887 – Albert A. Michelson and Edward W. Morley in their experiment do not detect the ether drift

- 1889 – Loránd Eötvös uses a torsion balance to test the weak equivalence principle to 1 part in one billion

- 1893 – Ernst Mach states Mach's principle; first constructive attack on the idea of Newtonian absolute space

- 1898 – Henri Poincaré states that simultaneity is relative

- 1899 – Hendrik Antoon Lorentz published Lorentz transformations

12.6   1900s

- 1904 – Henri Poincaré presents the principle of relativity for electromagnetism

- 1905 – Albert Einstein completes his theory of special relativity and states the law of mass-energy conservation: $E=mc^2$

- 1907 – Albert Einstein introduces the principle of equivalence of gravitation and inertia and uses it to predict the gravitational redshift

- 1915 – Albert Einstein completes his theory of general relativity. The new theory explains Mercury's strange motions that baffled Urbain Le Verrier.

- 1915 – Karl Schwarzschild publishes the Schwarzschild metric about a month after Einstein published his general theory of relativity. This was the first solution to the Einstein field equations other than the trivial flat space solution.

- 1916 – Albert Einstein shows that the field equations of general relativity admit wavelike solutions

- 1918 – J. Lense and Hans Thirring find the gravitomagnetic precession of gyroscopes in the equations of general relativity

- 1919 – Arthur Eddington leads a solar eclipse expedition which claims to detect gravitational deflection of light by the Sun

- 1921 – Theodor Kaluza demonstrates that a five-dimensional version of Einstein's equations unifies gravitation and electromagnetism

- 1937 – Fritz Zwicky states that galaxies could act as gravitational lenses

- 1937 – Albert Einstein, Leopold Infeld, and Banesh Hoffmann show that the geodesic equations of general relativity can be deduced from its field equations

12.6.1   1950s

- 1953: P. C. Vaidya Newtonian time in general relativity, Nature, **171**, p260.

- 1956: John Lighton Synge publishes the first relativity text emphasizing spacetime diagrams and geometrical methods,

- 1957: Felix A. E. Pirani uses Petrov classification to understand gravitational radiation,

- 1957: Richard Feynman introduces sticky bead argument,

- 1957 – John Wheeler discusses the breakdown of classical general relativity near singularities and the need for quantum gravity

- 1959: Pound–Rebka experiment, first precision test of gravitational redshift,

- 1959: Lluís Bel introduces Bel–Robinson tensor and the Bel decomposition of the Riemann tensor,

- 1959: Arthur Komar introduces the Komar mass,

- 1959: Richard Arnowitt, Stanley Deser and Charles W. Misner developed ADM formalism.

12.6.2   1960s

- 1960: Martin Kruskal and George Szekeres independently introduce the Kruskal–Szekeres coordinates for the Schwarzschild vacuum,

- 1960: Shapiro effect confirmed,

- 1960: Thomas Matthews and Allan R. Sandage associate 3C 48 with a point-like optical image, show radio source can be at most 15 light minutes in diameter,

- 1960: Carl H. Brans and Robert H. Dicke introduce Brans–Dicke theory, the first viable alternative theory with a clear physical motivation,

- 1960: Joseph Weber reports observation of gravitational waves (a claim now generally discounted),

- 1960: Ivor M. Robinson and Andrzej Trautman discover the Robinson-Trautman null dust solution[1]

- 1961: Pascual Jordan and Jürgen Ehlers develop the *kinematic decomposition* of a timelike congruence,

- 1960 – Robert Pound and Glen Rebka test the gravitational redshift predicted by the equivalence principle to approximately 1%

- 1962: Roger Penrose and Ezra T. Newman introduce the Newman–Penrose formalism,

- 1962: Ehlers and Wolfgang Kundt classify the symmetries of Pp-wave spacetimes,

- 1962: Joshua Goldberg and Rainer K. Sachs prove the Goldberg–Sachs theorem,

- 1962: Ehlers introduces Ehlers transformations, a new solution generating method,

- 1962: Cornelius Lanczos introduces the Lanczos potential for the Weyl tensor,

- 1962: Richard Arnowitt, Stanley Deser, and Charles W. Misner introduce the ADM reformulation and global hyperbolicity,

- 1962: Yvonne Choquet-Bruhat on Cauchy problem and global hyperbolicity,

- 1962: Istvan Ozsvath and Englbert Schücking rediscover the circularly polarized monochromomatic gravitational wave,

- 1962: Hans Adolph Buchdahl discovers Buchdahl's theorem,

- 1962: Hermann Bondi introduces Bondi mass,

- 1962 – Robert Dicke, Peter Roll, and R. Krotkov use a torsion fiber balance to test the weak equivalence principle to 2 parts in 100 billion

- 1963: Roy Kerr discovers the Kerr vacuum solution of Einstein's field equations,

- 1963: Redshifts of 3C 273 and other quasars show they are very distant; hence very luminous,

- 1963: Newman, T. Unti and L.A. Tamburino introduce the NUT vacuum solution,

- 1963: Roger Penrose introduces Penrose diagrams and Penrose limits,

- 1963: First Texas Symposium on Gravitational Astrophysics held in Dallas, 16–18 December,

- 1964: R. W. Sharp and Misner introduce the Misner–Sharp mass,

- 1964: M. A. Melvin discovers the Melvin electrovacuum solution (aka the *Melvin magnetic universe*),

- 1964 – Irwin Shapiro predicts a gravitational time delay of radiation travel as a test of general relativity

- 1965: Roger Penrose proves first of the singularity theorems,

- 1965: Newman and others discover the Kerr–Newman electrovacuum solution,

- 1965: Penrose discovers the structure of the light cones in gravitational plane wave spacetimes,

- 1965: Kerr and Alfred Schild introduce Kerr-Schild spacetimes,

- 1965: Subrahmanyan Chandrasekhar determines a stability criterion,

- 1965: Arno Penzias and Robert Wilson discover the cosmic microwave background radiation,

- 1965 – Joseph Weber puts the first Weber bar gravitational wave detector into operation

- 1966: Sachs and Ronald Kantowski discover the Kantowski-Sachs dust solution,

- 1967: Jocelyn Bell and Antony Hewish discover pulsars,

- 1967: Robert H. Boyer and R. W. Lindquist introduce Boyer–Lindquist coordinates for the Kerr vacuum,

- 1967: Bryce DeWitt publishes on canonical quantum gravity,

- 1967: Werner Israel proves the no-hair theorem,

- 1967: Kenneth Nordtvedt develops PPN formalism,

- 1967: Mendel Sachs publishes factorization of Einstein's field equations,

- 1967: Hans Stephani discovers the Stephani dust solution,

- 1968: F. J. Ernst discovers the Ernst equation,

- 1968: B. Kent Harrison discovers the Harrison transformation, a solution-generating method,

- 1968: Brandon Carter solves the geodesic equations for Kerr–Newmann electrovacuum,

- 1968: Hugo D. Wahlquist discovers the Wahlquist fluid,

- 1968 – Irwin Shapiro presents the first detection of the Shapiro delay

- 1968 – Kenneth Nordtvedt studies a possible violation of the weak equivalence principle for self-gravitating bodies and proposes a new test of the weak equivalence principle based on observing the relative motion of the Earth and Moon in the Sun's gravitational field

- 1969: William B. Bonnor introduces the Bonnor beam,

- 1969: Penrose proposes the (weak) cosmic censorship hypothesis and the Penrose process,

- 1969: Stephen W. Hawking proves area theorem for black holes,

- 1969: Misner introduces the mixmaster universe,

12.6.3   1970s

- 1970: Frank J. Zerilli derives the Zerilli equation,

- 1970: Vladimir A. Belinskiĭ, Isaak Markovich Khalatnikov, and Evgeny Lifshitz introduce the BKL conjecture,

- 1970: Chandrasekhar pushes on to 5/2 post-Newtonian order,

- 1970: Hawking and Penrose prove trapped surfaces must arise in black holes,

- 1970: the Kinnersley-Walker photon rocket,

- 1970: Peter Szekeres introduces colliding plane waves,

- 1971: Peter C. Aichelburg and Roman U. Sexl introduce the Aichelburg–Sexl ultraboost,

- 1971: Introduction of the Khan–Penrose vacuum, a simple explicit colliding plane wave spacetime,

- 1971: Robert H. Gowdy introduces the Gowdy vacuum solutions (cosmological models containing circulating gravitational waves),

- 1971: Cygnus X-1, the first solid black hole candidate, discovered by Uhuru satellite,

- 1971: William H. Press discovers black hole ringing by numerical simulation,

- 1971: Harrison and Estabrook algorithm for solving systems of PDEs,

- 1971: James W. York introduces conformal method generating initial data for ADM initial value formulation,

- 1971: Robert Geroch introduces Geroch group and a solution generating method,

- 1972: Jacob Bekenstein proposes that black holes have a non-decreasing entropy which can be identified with the area,

- 1972: Carter, Hawking and James M. Bardeen propose the four laws of black hole mechanics,

- 1972: Sachs introduces optical scalars and proves peeling theorem,

- 1972: Rainer Weiss proposes concept of interferometric gravitational wave detector,

- 1972: J. C. Hafele and R. E. Keating perform Hafele–Keating experiment,

- 1972: Richard H. Price studies gravitational collapse with numerical simulations,

- 1972: Saul Teukolsky derives the Teukolsky equation,

- 1972: Yakov B. Zel'dovich predicts the transmutation of electromagnetic and gravitational radiation,

- 1973: P. C. Vaidya and L. K. Patel introduce the Kerr–Vaidya null dust solution,

- 1973: Publication by Charles W. Misner, Kip S. Thorne and John A. Wheeler of the treatise *Gravitation*, the first modern textbook on general relativity,

- 1973: Publication by Stephen W. Hawking and George Ellis of the monograph *The Large Scale Structure of Space-Time*,

- 1973: Geroch introduces the GHP formalism,

- 1974: Russell Hulse and Joseph Hooton Taylor, Jr. discover the Hulse–Taylor binary pulsar,

- 1974: James W. York and Niall Ó Murchadha present the analysis of the initial value formulation and examine the stability of its solutions,

- 1974: R. O. Hansen introduces Hansen–Geroch multipole moments,

- 1974: Tullio Regge introduces the Regge calculus,

- 1974: Hawking discovers Hawking radiation,

- 1975: Chandrasekhar and Steven Detweiler compute quasinormal modes,

- 1975: Szekeres and D. A. Szafron discover the Szekeres–Szafron dust solutions,

- 1976: Penrose introduces Penrose limits (every null geodesic in a Lorentzian spacetime behaves like a plane wave),

- 1976 – Gravity Probe A experiment confirmed slowing the flow of time caused by gravity matching the predicted effects to an accuracy of about 70 parts per million.

- 1976 – Robert Vessot and Martin Levine use a hydrogen maser clock on a Scout D rocket to test the gravitational redshift predicted by the equivalence principle to approximately 0.007%

- 1978: Penrose introduces the notion of a *thunderbolt*,

- 1978: Belinskiĭ and Zakharov show how to solve Einstein's field equations using the inverse scattering transform; the first gravitational solitons,

- 1979: Richard Schoen and Shing-Tung Yau prove the positive mass theorem.

- 1979 – Dennis Walsh, Robert Carswell, and Ray Weymann discover the gravitationally lensed quasar Q0957+561

12.6.4   After 1980

- 1982 – Joseph Taylor and Joel Weisberg show that the rate of energy loss from the binary pulsar PSR B1913+16 agrees with that predicted by the general relativistic quadrupole formula to within 5%

- 2002 - First observations of Laser Interferometer Gravitational-Wave Observatory (LIGO), seeking direct detection of gravitational waves.

- 2007 – End of Gravity Probe B experiment.

12.7   See also

- Timeline of black hole physics

12.8   References

[1] "Spherical Gravitational Waves". Cdsads.u-strasbg.fr. Retrieved 2012-07-20.

12.9   External links

- Timeline of relativity and gravitation (Tomohiro Harada, Department of Physics, Rikkyo University)

- Timeline of General Relativity and Cosmology from 1905

Chapter 13

Timeline of nuclear fusion

See also: Fusion power § History of research

This **timeline of nuclear fusion** is an incomplete chronological summary of significant events in the study and use of nuclear fusion.

- **1920**

  - Based on F.W. Aston's measurements of the masses of low-mass elements and Einstein's discovery that $E=mc^2$, Arthur Eddington proposes that large amounts of energy released by fusing small nuclei together provides the energy source that powers the stars.[1]

- **1929**

  - Atkinson and Houtermans provide the first calculations of the rate of nuclear fusion in stars.[2]

- **1932**

  - Mark Oliphant discovered helium-3 and tritium, and that heavy hydrogen nuclei could be made to react with each other.

- **1939**

  - Hans Bethe shows how fusion powers the stars in work which won him the 1967 Nobel Prize in physics.
  - Peter Thonemann develops a detailed plan for a pinch device, but is told to do other work for his thesis.

- **1941**

  - Enrico Fermi proposed the idea of using a fission weapon to initiate nuclear fusion (still hypothetical) in a mass of hydrogen to Edward Teller. Teller became enthusiastic about the idea and worked on it (unsuccessfully) throughout the Manhattan Project.

- **1946**

  - George Paget Thomson and Moses Blackman patent the concept that would become known as the Z-pinch.

- **1947**

  - Thomson, Blackman, Thonemann, Cousins, Ware, Jim Tuck and other meet in Harwell to discuss the pinch approach and plan development.
  - First kiloampere plasma created by Cousins and Ware at the Imperial College, London, in a doughnut-shaped glass vacuum vessel. Plasmas are unstable and only last fractions of seconds.

- **1950**

  - The tokamak, a type of magnetic confinement fusion device, was proposed by soviet scientists Andrei Sakharov and Igor Tamm.

- **1951**

  - Edward Teller and Stanislaw Ulam at Los Alamos National Laboratory (LANL) develop the Teller-Ulam design for the thermonuclear weapon, allowing for the development of multi-megaton weapons.

  - Fusion work in the UK is classified after the Klaus Fuchs affair.

  - A press release from Argentina claims that their Huemul Project had produced controlled nuclear fusion. This prompted a wave of responses in other countries, especially the U.S.

  - Lyman Spitzer dismisses the Argentinian claims, but while thinking about it comes up with the stellarator concept. Funding is arranged under Project Matterhorn and develops into the Princeton Plasma Physics Laboratory.

  - Tuck introduces the British pinch work to LANL. He develops the Perhapsatron under the codename Project Sherwood. (Some people claim that the project was named Sherwood based on Friar Tuck. This claim is corroborated in a brief biographical sketch written by Tuck.[3])

Ivy Mike, the first thermonuclear weapon, in 1952

- **1952**

  - Ivy Mike shot of Operation Ivy, the first detonation of a thermonuclear weapon, yields 10.4 megatons of TNT out of a fusion fuel of liquid deuterium.

- Cousins and Ware build a larger toroidal pinch device in England, and demonstrated that instabilities in the plasma make pinch devices inherently unstable.

- **1953**

  - Pinch devices in the US and USSR attempted to take the reactions to fusion levels without worrying about stability. Both reported detections of neutrons, which were later explained as non-fusion in nature.

- **1954**

  - Construction of the ZETA device started at Harwell. It is the largest fusion device for some time.

  - Edward Teller gives a now-famous speech on plasma stability in magnetic bottles at the Princeton Gun Club. His work suggests that most magnetic bottles are inherently unstable.

- **1956**

  - Experimental research of tokamak systems started at Kurchatov Institute, Moscow by a group of Soviet scientists led by Lev Artsimovich.

  - Igor Kurchatov gives a talk at Harwell on pinch devices,[4] revealing for the first time that the USSR is also working on fusion. He details the problems they are seeing, mirroring those in the US and UK.

  - In the wake of the Kurchatov's speech, the US and UK begin to consider releasing their own data. Eventually they settle on a release prior to the 2nd Atoms for Peace conference in Geneva.

- **1957**

  - Initial results in ZETA appear to suggest the machine has successfully reached basic fusion temperatures. UK researchers start pressing for public release, while the US demurs.

- **1958**

  - The US and UK release large amounts of data in February, with the ZETA team claiming fusion. Other researchers, notably Artsimovich, are skeptical.

  - American, British and Soviet scientists began to share previously classified controlled fusion research as part of the Atoms for Peace conference in Geneva in September. It is the largest international scientific meeting to date. It becomes clear that basic pinch concepts are not successful.

- **1961**

  - The Soviet Union test the Tsar Bomba (50 megatons), the most powerful thermonuclear weapon ever.

- **1965** (approximate)

  - The 12-beam "4 pi laser" using ruby as the lasing medium is developed at Lawrence Livermore National Laboratory (LLNL) includes a gas-filled target chamber of about 20 centimeters in diameter.

- **1967**

  - Demonstration of Farnsworth-Hirsch Fusor appeared to generate neutrons in a nuclear reaction.

- **1968**

  - Results from the tokamak, a T-3 Soviet magnetic confinement device, which Igor Tamm and Andrei Sakharov had been working on, shows the temperatures in their machine to be over an order of magnitude higher than what was expected by the rest of the fusion community. The Western scientists visited the experiment and verified the high temperatures and confinement, sparking a wave of optimism for the prospects of the tokamak. It remains a dominant magnetic confinement device today, as well as development of new experiments.

- **1970**

- Kapchinskii and Teplyakov conceive the "The ion linear accelerator with space-uniform strong focusing". Demonstrated in 1979 at LANL, and named the radiofrequency quadrupole accelerator (RFQ). The concept increases the ion beam current that can be accelerated at low beta. This will be important for ICF drivers using high-energy heavy ions (HIF).

- **1972**

  - The first neodymium-doped glass (Nd:glass) laser for ICF research, the "Long Path laser" is completed at LLNL and is capable of delivering ~50 joules to a fusion target.

- **1973**

  - Design work on JET, the Joint European Torus, begins.

- **1974**

  - Taylor re-visited ZETA results of 1958 and explained that the quiet-period was in fact very interesting. This led to the development of reversed field pinch, now generalised as "self-organising plasmas", an ongoing line of research.

  - Construction completes and inertial confinement fusion experiments begin on the two beam Janus laser at the Lawrence Livermore National Laboratory.

  - On May 1, KMS Fusion (a private company founded by Kip Siegel) carried out the world's first successful laser-induced fusion in a deuterium-tritium pellet, the evidence for which was provided by neutron-sensitive nuclear emulsion detectors developed by Robert Hofstadter.

- **1975**

  - Heavy Ion Beams using mature high-energy accelerator technology are hailed as the elusive "brand-X" laser capable of driving fusion implosions for commercial power. The Livingston Curve, from Stanford SLAC Education Group, is modified to show the energy needed for fusion to occur. Experiments commence on the single beam LLNL Cyclops laser, testing new optical designs for future ICF lasers.

- **1976**

  - Workshop, called by the US-ERDA (now DoE) at the Claremont Hotel in Berkeley, CA for an ad-hoc two-week summer study. Fifty senior scientists from the major US ICF programs and accelerator laboratories participated, with program heads and Nobel laureates also attending. In the closing address, Dr. C. Martin Stickley, then Director of US-ERDA's Office of Inertial Fusion, announced the conclusion was "no show-stoppers" on the road to fusion energy.

  - The two beam Argus laser is completed at LLNL and experiments involving more advanced laser-target interactions commence.

- **1977**

  - The 20 beam Shiva laser at LLNL is completed, capable of delivering 10.2 kilojoules of infrared energy on target. At a price of $25 million and a size approaching that of a football field, the Shiva laser is the first of the "megalasers" at LLNL and brings the field of ICF research fully within the realm of "big science".

  - The JET project is given the go-ahead by the EC, choosing an ex-RAF airfield south east of Oxford, UK as its site.

- **1979**

  - LANL successfully demonstrates the radio frequency quadrupole accelerator (RFQ).

  - ANL and Hughes Research Laboratories demonstrate required ion source brightness with xenon beam at 1.5MeV.

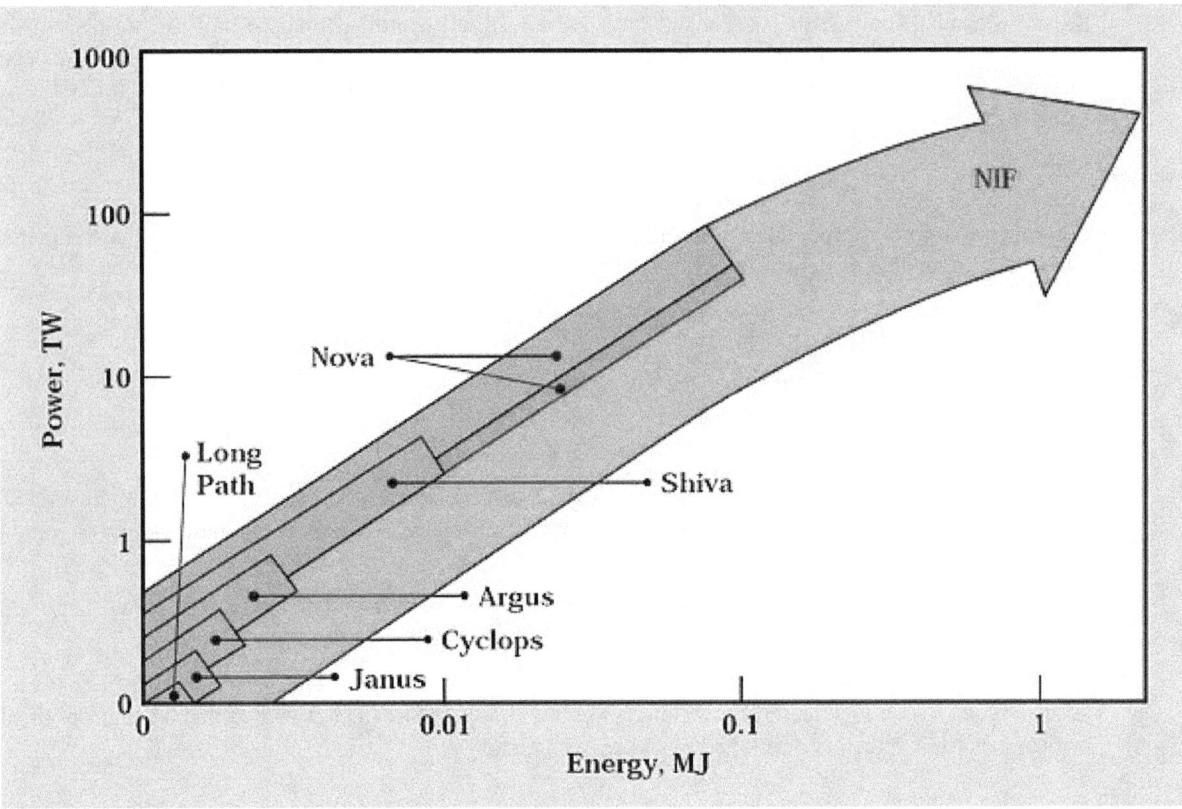

Progress in power and energy levels attainable by inertial confinement lasers has increased dramatically since the early 1970s.

- Foster Panel reports to US-DoE's Energy Research and Advisory Board that High-energy heavy ion fusion (HIF) is the "conservative approach" to fusion power. Listing HIF's advantages in his report, John Foster remarked: "…now that is kind of exciting." After DoE Office of Inertial Fusion completed review of programs, Director Gregory Canavan decides to accelerate the HIF effort.

- **1982**

  - HIBALL study by German and US institutions,[5] Garching uses the high repetition rate of the RF accelerator driver to serve four reactor chambers and first-wall protection using liquid lithium inside the chamber cavity.

  - Tore Supra construction starts at Cadarache, France. Its superconducting magnets will permit it to generate a strong permanent toroidal magnetic field.

- **1983**

  - JET, the largest is completed on time and on budget. First plasmas achieved.

  - The NOVETTE laser at LLNL comes on line and is used as a test bed for the next generation of ICF lasers, specifically the NOVA laser.

- **1984**

  - The huge 10 beam NOVA laser at LLNL is completed and switches on in December. NOVA would ultimately produce a maximum of 120 kilojoules of infrared laser light during a nanosecond pulse in a 1989 experiment.

- **1985**

  - National Academy of Sciences reviewed military ICF programs, noting HIF's major advantages clearly but averring that HIF was "supported primarily by other [than military] programs". The review of ICF by the National Academy of Sciences marked the trend with the observation: "The energy crisis is dormant for the time being." Energy becomes the sole purpose of heavy ion fusion.

- The Japanese tokamak, JT-60 completed. First plasmas achieved.

- **1988**

  - The T-15, Soviet tokamak with superconducting helium-cooled coils completed.

  - The Conceptual Design Activity for the International Thermonuclear Experimental Reactor (ITER), the successor to T-15, TFTR, JET and JT-60, begins. Participants include EURATOM, Japan, the Soviet Union and United States. It ended in 1990.

  - The first plasma produced at Tore Supra in April.[6]

- **1989**

  - On March 23, two Utah electrochemists, Stanley Pons and Martin Fleischmann, announced that they had achieved cold fusion: fusion reactions which could occur at room temperatures. However, they made their announcements before any peer review of their work was performed, and no subsequent experiments by other researchers revealed any evidence of fusion.

- **1990**

  - Decision to construct the National Ignition Facility "beamlet" laser at LLNL is made.

- **1991**

  - The START Tokamak fusion experiment begins in Culham. The experiment would eventually achieve a record beta (plasma pressure compared to magnetic field pressure) of 40% using a neutral beam injector. It was the first design that adapted the conventional toroidal fusion experiments into a tighter spherical design.

- **1992**

  - The Engineering Design Activity for the ITER starts with participants EURATOM, Japan, Russia and United States. It ended in 2001.

  - The United States and the Soviet Union cease nuclear weapons testing.

- **1993**

  - The TFTR tokamak at Princeton (PPPL) experiments with a 50% deuterium, 50% tritium mix, eventually producing as much as 10 megawatts of power from a controlled fusion reaction.

- **1994**

  - NIF Beamlet laser is completed and begins experiments validating the expected performance of NIF.

  - The USA declassifies information about indirectly driven (hohlraum) target design.

  - Comprehensive European-based study of HIF driver begins, centered at the Gesellshaft für Schwerionenforschung (GSI) and involving 14 laboratories, including USA and Russia. The Heavy Ion Driven Inertial Fusion (HIDIF) study will be completed in 1997.

- **1996**

  - A record is reached at Tore Supra: a plasma duration of two minutes with a current of almost 1 million amperes driven non-inductively by 2.3 MW of lower hybrid frequency waves (i.e. 280 MJ of injected and extracted energy). This result was possible due to the actively cooled plasma-facing components installed in the machine.[7]

- **1997**

  - The JET tokamak in the UK produces 16 MW of fusion power - the current world record for fusion power. Four megawatts of alpha particle self-heating was achieved.

  - LLNL study compared projected costs of power from ICF and other fusion approaches to the projected future costs of existing energy sources. **Groundbreaking ceremony held for the National Ignition Facility (NIF).

- **1998**

  - The JT-60 tokamak in Japan produced a high performance reversed shear plasma with the equivalent fusion amplification factor Q_{eq} of 1.25 - the current world record of Q, fusion energy gain factor.
  - Results of European-based study of heavy ion driven fusion power system (HIDIF, GSI-98-06) incorporates telescoping beams of multiple isotopic species. This technique multiplies the 6-D phase space usable for the design of HIF drivers.

- **1999**

  - The United States withdraws from the ITER project.
  - The START experiment is succeeded by MAST.

- **2001**

  - Building construction for the immense 192-beam 500-terawatt NIF project is completed and construction of laser beam-lines and target bay diagnostics commences, expecting to take its first full system shot in 2010.
  - Negotiations on the Joint Implementation of ITER begin between Canada, countries represented by the European Union, Japan and Russia.

- **2002**

  - Claims and counter-claims are published regarding bubble fusion, in which a table-top apparatus was reported as producing small-scale fusion in a liquid undergoing acoustic cavitation. Like cold fusion (see 1989), it is later dismissed.
  - European Union proposes Cadarache in France and Vandellos in Spain as candidate sites for ITER while Japan proposes Rokkasho.

- **2003**

  - The United States rejoins the ITER project with China and Republic of Korea also joining. Canada withdraws.
  - Cadarache in France is selected as the European Candidate Site for ITER.
  - Sandia National Laboratories began fusion experiments in the Z machine.

- **2004**

  - The United States drops its own projects, recognising an inability to match EU progress (Fusion Ignition Research Experiment (FIRE)), and focuses resources on ITER.

- **2005**

  - Following final negotiations between the EU and Japan, ITER chooses Cadarache over Rokkasho for the site of the reactor. In concession, Japan is able to host the related materials research facility and granted rights to fill 20% of the project's research posts while providing 10% of the funding.
  - The NIF fires its first bundle of eight beams achieving the highest ever energy laser pulse of 152.8 kJ (infrared).

- **2006**

  - China's EAST test reactor is completed, the first tokamak experiment to use superconducting magnets to generate both the toroidal and poloidal fields.

- **2009**

  - Construction of the NIF reported as complete.
  - Ricardo Betti, the third Under Secretary, responsible for Nuclear Energy, testifies before Congress: "IFE [ICF for energy production] has no home".
  - Fusion Power Corporation files patent application for "Single Pass RF Driver" a RF Accelerator Driven HIF Process and Method.

- **2010**

    - HIF-2010 Symposium in Darmstadt Germany. Robert J Burke presented on Single Pass HIF and Charles Helsley made a presentation on the commercialization of HIF within the decade.

- **2011**

    - May 23–26, Workshop for Accelerators for Heavy Ion Fusion at Lawrence Berkeley National Laboratory, presentation by Robert J. Burke on "Single Pass Heavy Ion Fusion". The Accelerator Working Group publishes recommendations supporting moving RF accelerator Driven HIF toward commercialization.

- **2012**

    - Stephen Slutz & Roger Vesey of Sandia National Labs publish a paper in Physical Review Letters presenting a computer simulation of the MagLIF concept showing it can produce high gain. According to the simulation, a 70 Mega Amp Z-pinch facility in combination with a Laser may be able to produce spectacular energy return of 1000 times the expended energy. A 60 MA facility would produce a 100x yield.[8]

    - JET announces a major breakthrough in controlling instabilities in a fusion plasma.

    - In August Robert J. Burke presents updates to the SPRFD HIF process[9] and Charles Helsley presents the Economics of SPRFD[10] at the 19th International HIF Symposium at Berkeley, California. Industry was there in support of ion generation for SPRFD.

    - Fusion Power Corporation SPRFD patent allowed in Russia.

- **2013**

    - EAST tokamak test reactor achieves a record confinement time of 30 seconds for plasma in the high-confinement mode (H-mode), thanks to improvements in heat dispersal from tokamak walls. This is an improvement of an order of magnitude with respect to state-of-the-art reactors.

- **2014**

    - US Scientists successfully generate more energy from fusion reactions than they put into the nuclear fuel at NIF.[11]

13.1 Notes

[1] Eddington, A. S. (October 1920). "The internal constitution of the stars". *The Observatory* **43**: 341–358. Bibcode:1920Obs.... Retrieved 20 July 2015. 43..341E.

[2] Atkinson, R. d E.; Houtermans, F. G. (1929). "Zur Frage der Aufbaumöglichkeit der Elemente in Sternen". *Zeitschrift für Physik* **54** (9-10): 656–665. doi:10.1007/BF01341595. Retrieved 20 July 2015.

[3] ...the first money to be allocated [to controlled nuclear research] happened to be for Tuck, and was diverted from Project Lincoln, in the Hood Laboratory. The coincidence of names prompted the well-known cover name "Project Sherwood". James L. Tuck, "Curriculum Vita and Autobiography", declassified document from Los Alamos National Laboratory (1974), reproduced with permission. Archived 9 February 2012.

[4] "Lecture of I.V. Kurchatov at Harwell", from the address of I.V. Kurchatov: "On the possibility of producing thermonuclear reactions in a gas discharge" at Harwell on 25 April 1956. Archived 20 July 2015.

[5] ...Gesellschaft für Schwerionenforschung; Institut für Plasmaphysik, Garching; Kernforschungszentrum Karlsruhe, University of Wisconsin, Madison; Max-Planck-Institut für Quantenoptik

[6] "The Tore Supra Tokamak". CEA. Archived from the original on 11 February 2011. Retrieved 20 July 2015.

[7] http://www-drfc.cea.fr/gb/cea/ts/ts.htm

[8] Slutz, Stephen A.; Vesey, Roger A. (12 January 2012). "High-Gain Magnetized Inertial Fusion". *Phys. Rev. Lett.* **108** (2). doi:10.1103/PhysRevLett.108.025003.

[9] Burke, Robert (1 January 2014). "The Single Pass RF Driver: Final beam compression". *Nuclear Instruments and Methods in Physics Research Section A: Accelerators, Spectrometers, Detectors and Associated Equipment* **733**: 158–167. doi:10.1016/j.nima.

[10] Helsley, Charles E.; Burke, Robert J. "Economic viability of large-scale fusion systems". *Nuclear Instruments and Methods in Physics Research Section A: Accelerators, Spectrometers, Detectors and Associated Equipment* **733**: 51–56. doi:10.1016/j.nima.

[11] Herrmann, Mark (20 February 2014). "Plasma physics: A promising advance in nuclear fusion". *Nature* **506**. doi:10.1038/nature

13.2   External links

- Fusion experiments from the British Science Museum

- International Fusion Research Council, Status report on fusion research, *Nuclear Fusion* **45**:10A, October 2005.

Chapter 14

Timeline of theoretical physics

The **Timeline of theoretical physics** lists key events by century.

14.1   17th century

- 1687 - Newton: Laws of Motion and Law of Gravity[1]

14.2   18th century

- 1782 ? - Lavoisier: Conservation of matter
- 1785 - Coulomb: Inverse-square law for electric charges confirmed[2]

14.3   19th century

- 1801 - Young: Wave theory of light
- 1803 - Dalton: Atomic theory of matter
- 1806 - Young: Kinetic energy
- 1814 - Fresnel: Wave theory of light, interference
- 1820 - Ampère, Biot, & Savart: Evidence for electromagnetic interactions
- 1824 - Sadi Carnot: Ideal gas cycle analysis, internal combustion engine
- 1827 - Ohm: Electrical resistance, etc.
- 1838 - Michael Faraday: Lines of force, Fields
- 1838 - Weber: Earth's magnetic field
- 1842-43 - Kelvin & Mayer: Conservation of energy
- 1842 - Kelvin: Doppler effect
- 1845 - Faraday: Faraday Rotation (light and electromagnetic)
- 1847 - Helmholtz & Joule: Conservation of Energy 2

- 1850-51 - Kelvin & Clausius: Second law of thermodynamics

- 1857-59 - Clausius & Maxwell: Kinetic theory

- 1861 - Kirchhoff: Black body

- 1863 - Clausius: Entropy

- 1864 - Maxwell: Dynamical Theory of the Electromagnetic Field

- 1867 - Maxwell: Dynamic theory of gases

- 1871-89 - Boltzmann & Gibbs: Statistical mechanics

- 1884 - Boltzmann derives Stefan radiation law

- 1887 - Hertz: Electromagnetic waves

- 1893 - Wien: Radiation law

- 1895 - Röntgen: X-rays

- 1896 - Becquerel: Radioactivity

- 1897 - Thomson: Electron

14.4   20th century

- 1900 - Planck: Formula for black-body radiation

- 1905 - Einstein: Special relativity, Photoelectric effect & Brownian motion

- 1911 - Rutherford: Discovery of the atomic nucleus, Kamerlingh Onnes: Superconductivity & equivalence principle

- 1913 - Bohr: Bohr model of the atom

- 1916 - Einstein: General relativity

- 1919 - Light Bending confirmed

- 1922 - Friedmann proposes expanding universe

- 1923 - Stern–Gerlach experiment, Matter waves, galaxies & particle nature of photons confirmed

- 1925-27 - Quantum mechanics

- 1925 - Stellar structure understood

- 1927 - Lemaître: Big Bang

- 1928 - Dirac: Antimatter predicted

- 1929 - Hubble: Expansion of universe confirmed

- 1932 - Anderson: Antimatter discovered & Chadwick: Neutron discovered

- 1933 - Invention of the electron microscope by Ernst Ruska

- 1937 - Muon discovered by Anderson & Neddermeyer

- 1938 - Superfluidity discovered & Energy production in stars understood

- 1939 - Uranium fission discoveredo

- 1944 - Theory of magnetism in 2D: Ising model

- 1947 - Pion discovered

- 1948 - Quantum electrodynamics

- 1948 - Invention of the maser and laser by Charles Townes

- 1956 - Electron neutrino discovered

- 1956-57 - Parity found violated

- 1957 - Superconductivity explained

- 1959-60 - Role of topology in quantum physics predicted and Confirmed

- 1962 - SU(3) theory of strong interactions & muon neutrino found

- 1963 - Murray Gell-Mann & George Zweig: Quarks predicted

- 1967 - Unification of weak and electromagnetic interactions, Solar neutrino Problem found & Pulsars (neutron stars) discovered

- 1968 - Experimental evidence for quarks found

- 1968 - Vera Rubin: Dark matter theories

- 1970-73 - Standard Model of elementary particles invented

- 1971 - Helium 3 Superfluidity

- 1974 - Black hole radiation predicted, renormalization group & charmed quark found

- 1975 - Tau lepton found

- 1977 - Bottom quark found

- 1980 - Quantum Hall effect

- 1981 - Theory of Cosmic Inflation proposed

- 1982 - Fractional quantum Hall effect

- 1995 - Wolfgang Ketterle: Bose–Einstein condensate found

- 1995 - Top quark found

- 1998 - Accelerating expansion of Universe found

- 1999 - Lene Vestergaard Hau: Slow light experimentally demonstrated

14.5   21st century

- 2000 - Tau neutrino found

- 2003 - WMAP Observations of cosmic microwave background

- 2012 - Higgs boson found

- 2014 - Gravitational waves detected from Cosmic microwave background

14.6   See also

- Physics

- Timeline of developments in theoretical physics

14.7   References

[1] American Heritage Dictionary (January 2005). *The American Heritage Science Dictionary*. Houghton Mifflin Harcourt. p. 428. ISBN 978-0-618-45504-1.

[2] John L. Heilbron (14 February 2003). *The Oxford Companion to the History of Modern Science*. Oxford University Press. p. 235. ISBN 978-0-19-974376-6.

Chapter 15

Timeline of developments in theoretical physics

This page lists important developments in theoretical physics that have either been experimentally confirmed or significantly influence current thinking in modern physics.

Discovery	Date	Literature
Kepler's laws of planetary motion	1609, 1619	
The Galilean principle	1632	
Newton's laws of motion	1687	
Newton's law of universal gravitation	1687	
Laws of thermodynamics	1824, 1873, 1930s	
Coulomb's law	1785	
Biot–Savart law	1820	
Ampere's Law	1826	
Faraday's law of induction	1831	
Boltzmann equation	1872	
Maxwell's equations and electromagnetic waves	1873	
Lorentz-Fitzgerald contraction	1889, 1892	
Planck proposes quanta solution to radiation ultraviolet catastrophe	1900	
Einstein proposes the photon	1905	
Special relativity	1905	
General relativity	1916	
Schwarzschild metric	1916	
Kaluza-Klein models		
FRW Metric	1922, 1935, 1937	
Chandrasekhar limit	1935	
J. J. Thompson model		
Rutherford model	1912	
Bohr model	1913	
Bose–Einstein statistics	1924	
De Broglie wave	1924	
Heisenberg Uncertainty Principle	1927	
Heisenberg Matrix Mechanics	1925	
Fermi-Dirac Statistics	1926	
Schrödinger equation	1926	
Born Interpretation	1927	
Dirac equation	1927	
Dirac antiparticle	1930	
S-Matrix		
Feynman Path integral	1941	
Feynman rules and diagrams	1948	
Quantum electrodynamics	1948	
Yang–Mills theory		
Spontaneous symmetry breaking		
Electroweak unification		
Renormalizability of gauge theories		
Renormalization group		
Veneziano model		
QCD, quarks and gluons		
Supersymmetry		
Supergravity		
Black Hole Entropy	1972	
Hawking radiation	1974	
Cosmic inflation		
BRST symmetry		
Skyrmions		
Quantum computers		
Bosonic strings		
NSR strings		
Green–Schwarz superstrings		
Heterotic strings		
T-duality		
Loop quantum gravity		
SUSY dark matter		
Dark energy as cosmological constant		
Matrix models/M-theory	1994, 1996, 1997	
Randall-Sundrum models		
Mirror symmetry		
AdS/CFT and the holographic principle		

Chapter 16

Timeline of quantum mechanics

This **timeline of quantum mechanics** shows the key steps, precursors and contributors to the development of quantum mechanics, quantum field theories and quantum chemistry.[1][2]

16.1   19th century

- 1859 – Kirchhoff introduces the concept of a blackbody and proves that its emission spectrum depends only on its temperature.[1]

- 1860–1900 – Ludwig Eduard Boltzmann, James Clerk Maxwell and others develop the theory of statistical mechanics. Boltzmann argues that entropy is a measure of disorder.[1]

- 1877 – Boltzmann suggests that the energy levels of a physical system could be discrete based on statistical mechanics and mathematical arguments; also produces the first circle diagram representation, or atomic model of a molecule (such as an iodine gas molecule) in terms of the overlapping terms α and β, later (in 1928) called molecular orbitals, of the constituting atoms.

- 1887 – Heinrich Hertz discovers the photoelectric effect, shown by Einstein in 1905 to involve *quanta* of light.

- 1888 – Hertz demonstrates experimentally that electromagnetic waves exist, as predicted by Maxwell.[1]

- 1888 – Johannes Rydberg modifies the Balmer formula to include all spectral series of lines for the hydrogen atom, producing the Rydberg formula which is employed later by Niels Bohr and others to verify Bohr's first quantum model of the atom.

- 1895 – Wilhelm Conrad Röntgen discovers X-rays in experiments with electron beams in plasma.[1]

- 1896 – Antoine Henri Becquerel accidentally discovers radioactivity while investigating the work of Wilhelm Conrad Röntgen; he finds that uranium salts emit radiation that resembled Röntgen's X-rays in their penetrating power. In one experiment, Becquerel wraps a sample of a phosphorescent substance, potassium uranyl sulfate, in photographic plates surrounded by very thick black paper in preparation for an experiment with bright sunlight; then, to his surprise, the photographic plates are already exposed before the experiment starts, showing a projected image of his sample.[1][3]

- 1896 – Pieter Zeeman first observes the Zeeman splitting effect by passing the light emitted by hydrogen through a magnetic field.

- 1896–1897 Marie Curie (née Skłodowska, Becquerel's doctoral student) investigates uranium salt samples using a very sensitive electrometer device that was invented 15 years before by her husband and his brother Jacques Curie to measure electrical charge. She discovers that rays emitted by the uranium salt samples make the surrounding air electrically conductive, and measures the emitted rays' intensity. In April 1898, through a systematic search of

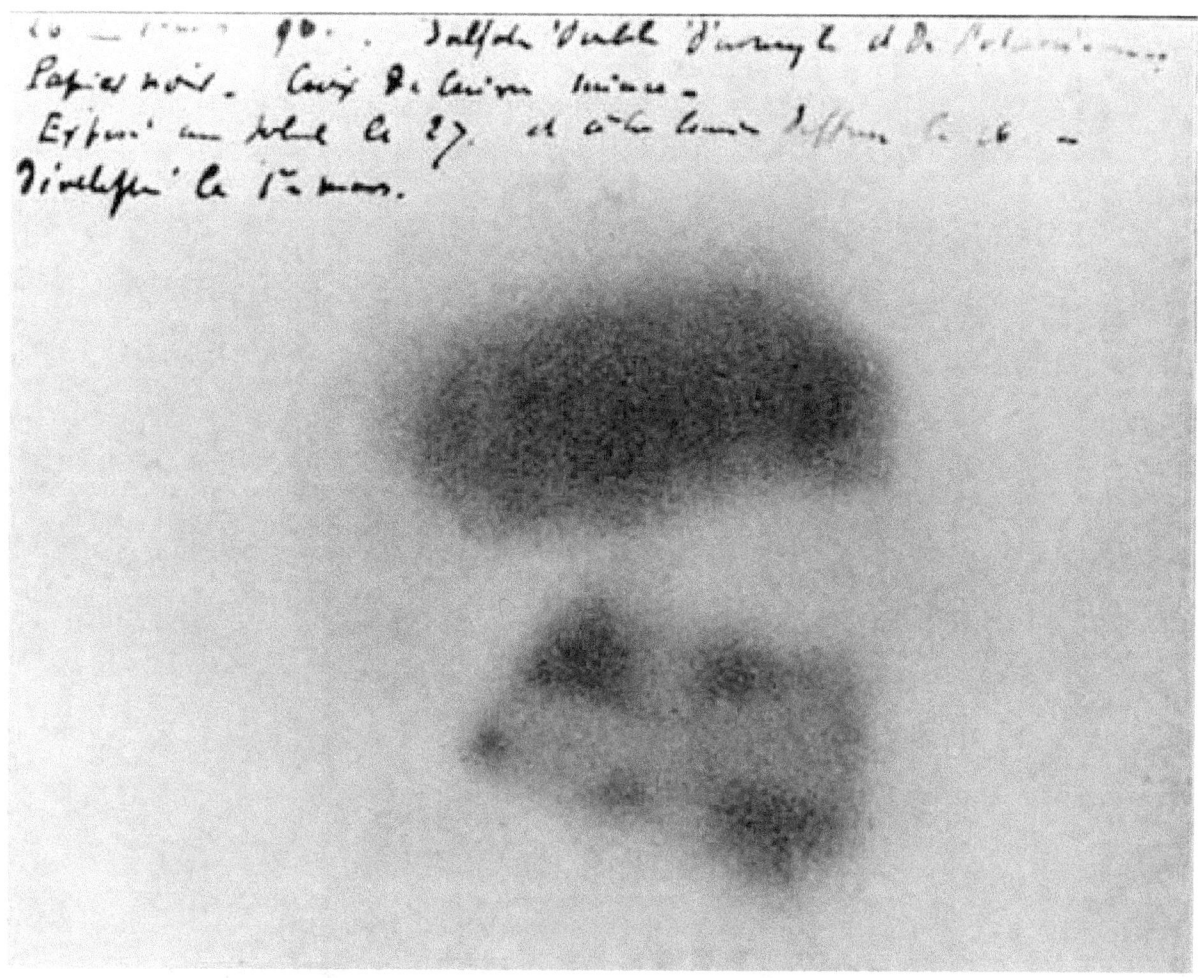

Image of Becquerel's photographic plate which has been fogged by exposure to radiation from a uranium salt. The shadow of a metal Maltese Cross placed between the plate and the uranium salt is clearly visible.

substances, she finds that thorium compounds, like those of uranium, emitted "Becquerel rays", thus preceding the work of Frederick Soddy and Ernest Rutherford on the nuclear decay of thorium to radium by three years.[4]

- 1897 – Ivan Borgman demonstrates that X-rays and radioactive materials induce thermoluminescence.

- 1899 to 1903 – Ernest Rutherford, 1st Baron, Lord Rutherford of Nelson, of Cambridge, OM, FRS: During the investigation of radioactivity he coins the terms alpha and beta rays in 1899 to describe the two distinct types of radiation emitted by thorium and uranium salts. Ernest Rutherford is joined at McGill University in 1900 by Frederick Soddy and together they discover nuclear transmutation when they find in 1902 that radioactive thorium is converting itself into radium through a process of nuclear decay and a gas (later found to be ^4_2He); they report their interpretation of radioactivity in 1903.[5] Sir Ernest Rutherford becomes known as the "father of nuclear physics": with his nuclear atom model of 1911 he leads the exploration of nuclear physics.[6]

16.2   20th century

16.2.1   1900–1909

- 1900 – To explain black-body radiation (1862), Max Planck suggests that electromagnetic energy could only be emitted in quantized form, i.e. the energy could only be a multiple of an elementary unit $E = h\nu$, where h is Planck's

Einstein, in 1905, when he wrote the Annus Mirabilis papers

constant and ν is the frequency of the radiation.

- 1902 – To explain the octet rule (1893), Gilbert N. Lewis develops the "cubical atom" theory in which electrons in the form of dots are positioned at the corner of a cube. Predicts that single, double, or triple "bonds" result when two atoms are held together by multiple pairs of electrons (one pair for each bond) located between the two atoms.

- 1903 – Antoine Becquerel, Pierre Curie and Marie Curie share the 1903 Nobel Prize in Physics for their work on spontaneous radioactivity.

- 1904 – Richard Abegg notes the pattern that the numerical difference between the maximum positive valence, such as +6 for H_2SO_4, and the maximum negative valence, such as −2 for H_2S, of an element tends to be eight (Abegg's rule).

- 1905 – Albert Einstein explains the photoelectric effect (reported in 1887 by Heinrich Hertz), i.e. that shining light on certain materials can function to eject electrons from the material. He postulates, as based on Planck's quantum hypothesis (1900), that light itself consists of individual quantum particles (photons).

- 1905 – Einstein explains the effects of Brownian motion as caused by the kinetic energy (i.e., movement) of atoms, which was subsequently, experimentally verified by Jean Baptiste Perrin, thereby settling the century-long dispute about the validity of John Dalton's atomic theory.

- 1905 – Einstein publishes his Special Theory of Relativity.

- 1905 – Einstein theoretically derives the equivalence of matter and energy.

- 1907 to 1917 – Ernest Rutherford: To test his *planetary* model of 1904, later known as the Rutherford model, he sent a beam of positively charged alpha particles onto a gold foil and noticed that some bounced back, thus showing that an atom has a small-sized positively charged atomic nucleus at its center. However, he received in 1908 the Nobel Prize in Chemistry "for his investigations into the disintegration of the elements, and the chemistry of radioactive substances",[7] which followed on the work of Marie Curie, not for his planetary model of the atom; he is also widely credited with first "splitting the atom" in 1917. In 1911 Ernest Rutherford explained the Geiger–Marsden experiment by invoking a nuclear atom model and derived the Rutherford cross section.

- 1909 – Geoffrey Ingram Taylor demonstrates that interference patterns of light were generated even when the light energy introduced consisted of only one photon. This discovery of the wave–particle duality of matter and energy is fundamental to the later development of quantum field theory.

- 1909 and 1916 – Einstein shows that, if Planck's law of black-body radiation is accepted, the energy quanta must also carry momentum $p = h / \lambda$, making them full-fledged particles.

16.2.2   1910–1919

- 1911 – Lise Meitner and Otto Hahn perform an experiment that shows that the energies of electrons emitted by beta decay had a continuous rather than discrete spectrum. This is in apparent contradiction to the law of conservation of energy, as it appeared that energy was lost in the beta decay process. A second problem is that the spin of the Nitrogen-14 atom was 1, in contradiction to the Rutherford prediction of ½. These anomalies are later explained by the discoveries of the neutrino and the neutron.

- 1911 – Ştefan Procopiu performs experiments in which he determines the correct value of electron's magnetic dipole moment, $\mu B = 9.27 \times 10^{-21}$ erg·Oe^{-1} (in 1913 he is also able to calculate a theoretical value of the Bohr magneton based on Planck's quantum theory).

- 1912 – Victor Hess discovers the existence of cosmic radiation.

- 1912 – Henri Poincaré publishes an influential mathematical argument in support of the essential nature of energy quanta.[8][9]

- 1913 – Robert Andrews Millikan publishes the results of his "oil drop" experiment, in which he precisely determines the electric charge of the electron. Determination of the fundamental unit of electric charge makes it possible to calculate the Avogadro constant (which is the number of atoms or molecules in one mole of any substance) and thereby to determine the atomic weight of the atoms of each element.

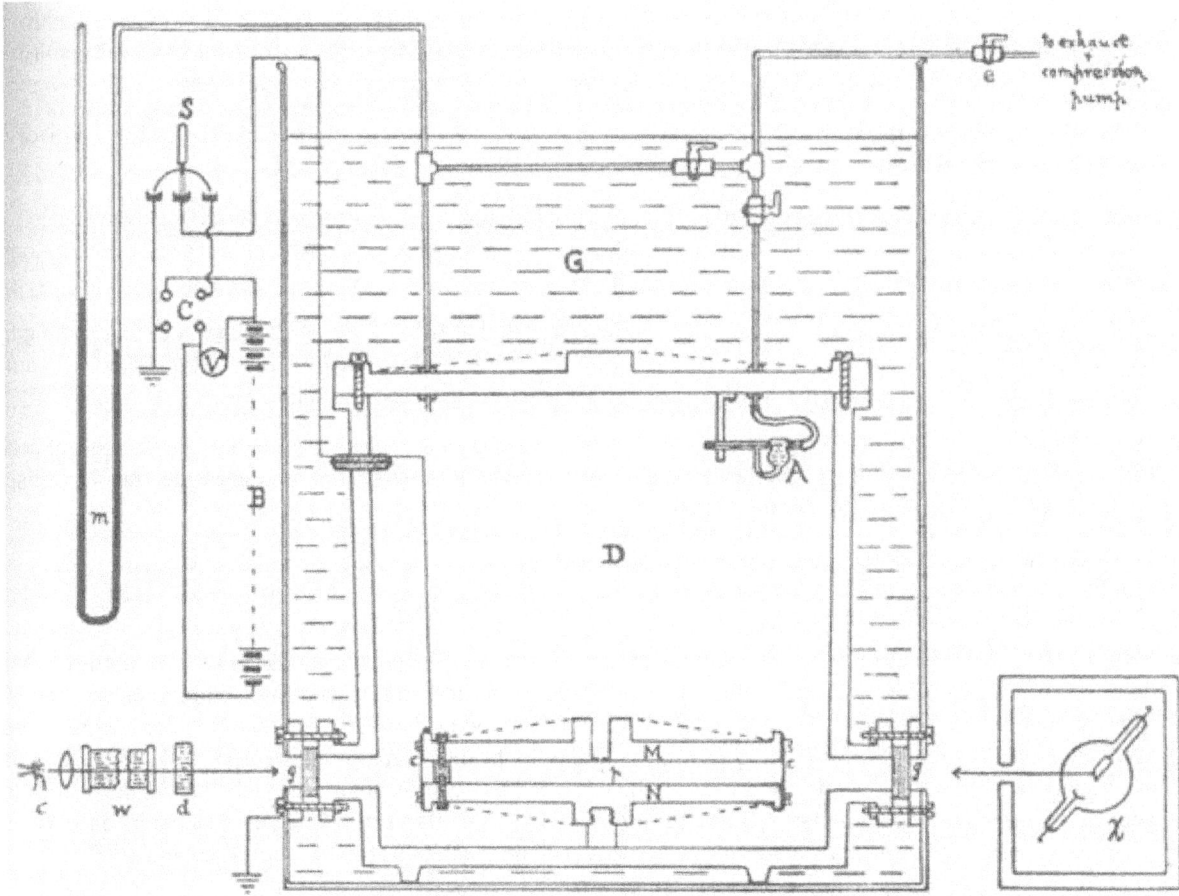

A schematic diagram of the apparatus for Millikan's refined oil drop experiment.

- 1913 – Ștefan Procopiu publishes a theoretical paper with the correct value of the electron's magnetic dipole moment μB.[10]

- 1913 – Niels Bohr obtains theoretically the value of the electron's magnetic dipole moment μB as a consequence of his atom model

- 1913 – Johannes Stark and Antonino Lo Surdo independently discover the shifting and splitting of the spectral lines of atoms and molecules due to the presence of the light source in an external static electric field.

- 1913 – To explain the Rydberg formula (1888), which correctly modeled the light emission spectra of atomic hydrogen, Bohr hypothesizes that negatively charged electrons revolve around a positively charged nucleus at certain fixed "quantum" distances and that each of these "spherical orbits" has a specific energy associated with it such that electron movements between orbits requires "quantum" emissions or absorptions of energy.

- 1914 – James Franck and Gustav Hertz report their experiment on electron collisions with mercury atoms, which provides a new test of Bohr's quantized model of atomic energy levels.[11]

- 1915 – Einstein first presents to the Prussian Academy of Science what are now known as the Einstein field equations. These equations specify how the geometry of space and time is influenced by whatever matter is present, and form the core of Einstein's General Theory of Relativity. Although this theory is not directly applicable to quantum mechanics, theorists of quantum gravity seek to reconcile them.

- 1916 – Paul Epstein[12] and Karl Schwarzschild,[13] working independently, derive equations for the linear and quadratic Stark effect in hydrogen.

- 1916 – To account for the Zeeman effect (1896), i.e. that atomic absorption or emission spectral lines change when the light source is subjected to a magnetic field, Arnold Sommerfeld suggests there might be "elliptical orbits" in atoms in addition to spherical orbits.

- 1918 – Sir Ernest Rutherford notices that, when alpha particles are shot into nitrogen gas, his scintillation detectors shows the signatures of hydrogen nuclei. Rutherford determines that the only place this hydrogen could have come from was the nitrogen, and therefore nitrogen must contain hydrogen nuclei. He thus suggests that the hydrogen nucleus, which is known to have an atomic number of *1*, is an elementary particle, which he decides must be the protons hypothesized by Eugen Goldstein.

- 1919 – Building on the work of Lewis (1916), Irving Langmuir coins the term "covalence" and postulates that coordinate covalent bonds occur when two electrons of a pair of atoms come from both atoms and are equally shared by them, thus explaining the fundamental nature of chemical bonding and molecular chemistry.

16.2.3   1920–1929

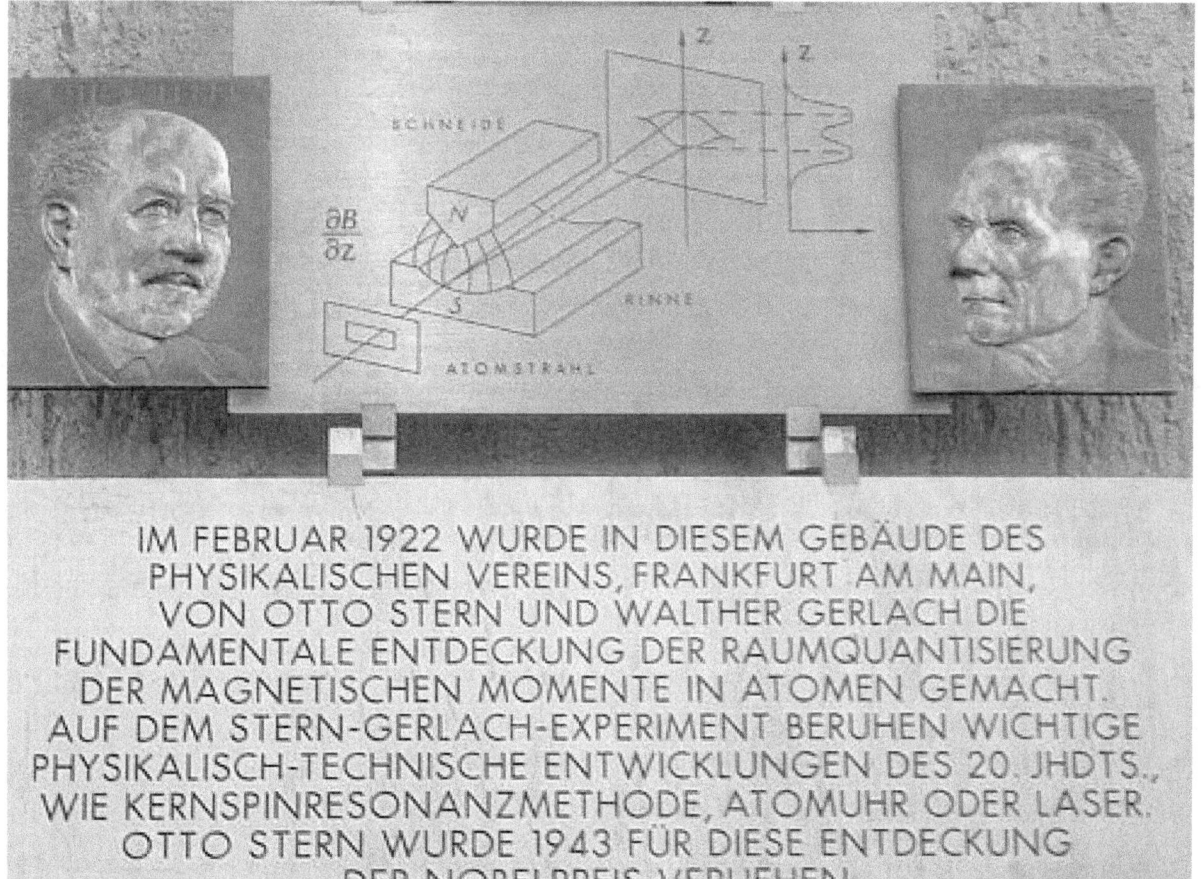

A plaque at the University of Frankfurt commemorating the Stern–Gerlach experiment.

- 1921–1922 – Frederick Soddy receives the Nobel Prize for 1921 in Chemistry one year later, in 1922, "for his contributions to our knowledge of the chemistry of radioactive substances, and his investigations into the origin and nature of isotopes"; he writes in his Nobel Lecture of 1922: "The interpretation of radioactivity which was published in 1903 by Sir Ernest Rutherford and myself ascribed the phenomena to the spontaneous disintegration of the atoms of the radio-element, whereby a part of the original atom was violently ejected as a radiant particle, and the remainder formed a totally new kind of atom with a distinct chemical and physical character."

- 1922 – Arthur Compton finds that X-ray wavelengths increase due to scattering of the radiant energy by free electrons. The scattered quanta have less energy than the quanta of the original ray. This discovery, known as the *Compton effect* or Compton scattering, demonstrates the particle concept of electromagnetic radiation.

- 1922 – Otto Stern and Walther Gerlach perform the Stern–Gerlach experiment, which detects discrete values of angular momentum for atoms in the ground state passing through an inhomogeneous magnetic field leading to the discovery of the spin of the electron.

- 1922 – Bohr updates his model of the atom to better explain the properties of the periodic table by assuming that certain numbers of electrons (for example 2, 8 and 18) corresponded to stable "closed shells", presaging orbital theory.

- 1923 – Pierre Auger discovers the Auger effect, where filling the inner-shell vacancy of an atom is accompanied by the emission of an electron from the same atom.

- 1923 – Louis de Broglie extends wave–particle duality to particles, postulating that electrons in motion are associated with waves. He predicts that the wavelengths are given by Planck's constant h divided by the momentum of the $mv = p$ of the electron: $\lambda = h \, / \, mv = h \, / \, p$.[1]

- 1923 – Gilbert N. Lewis creates the theory of Lewis acids and bases based on the properties of electrons in molecules, defining an acid as accepting an electron lone pair from a base.

- 1924 – Satyendra Nath Bose explains Planck's law using a new statistical law that governs bosons, and Einstein generalizes it to predict Bose–Einstein condensate. The theory becomes known as Bose–Einstein statistics.[1]

- 1924 – Wolfgang Pauli outlines the "Pauli exclusion principle" which states that no two identical fermions may occupy the same quantum state simultaneously, a fact that explains many features of the periodic table.[1]

- 1925 – George Uhlenbeck and Samuel Goudsmit postulate the existence of electron spin.[1]

- 1925 – Friedrich Hund outlines Hund's rule of Maximum Multiplicity which states that when electrons are added successively to an atom as many levels or orbits are singly occupied as possible before any pairing of electrons with opposite spin occurs and made the distinction that the inner electrons in molecules remained in atomic orbitals and only the valence electrons needed to be in molecular orbitals involving both nuclei.

- 1925 – Werner Heisenberg, Max Born, and Pascual Jordan develops the matrix mechanics formulation of Quantum Mechanics.[1]

- 1926 – Lewis coins the term photon in a letter to the scientific journal Nature, which he derives from the Greek word for light, φως (transliterated phôs).[14]

- 1926 – Oskar Klein and Walter Gordon state their relativistic quantum wave equation, later called the Klein–Gordon equation.

- 1926 – Enrico Fermi discovers the spin-statistics theorem connection.

- 1926 – Paul Dirac introduces Fermi–Dirac statistics.

- 1926 – Erwin Schrödinger uses De Broglie's electron wave postulate (1924) to develop a "wave equation" that represents mathematically the distribution of a charge of an electron distributed through space, being spherically symmetric or prominent in certain directions, i.e. directed valence bonds, which gives the correct values for spectral lines of the hydrogen atom; also introduces the Hamiltonian operator in quantum mechanics.

- 1926 – Paul Epstein reconsiders the linear and quadratic Stark effect from the point of view of the new quantum theory, using the equations of Schrödinger and others. The derived equations for the line intensities are a decided improvement over previous results obtained by Hans Kramers.[15]

- 1926 to 1932 – John von Neumann lays the mathematical foundations of Quantum Mechanics in terms of Hermitian operators on Hilbert spaces, subsequently published in 1932 as a basic textbook of quantum mechanics.[1][16]

- 1927 – Werner Heisenberg formulates the quantum uncertainty principle.[1]

- 1927 – Max Born develops the Copenhagen interpretation of the probabilistic nature of wavefunctions.

- 1927 – Born and J. Robert Oppenheimer introduce the Born–Oppenheimer approximation, which allows the quick approximation of the energy and wavefunctions of smaller molecules.

- 1927 – Walter Heitler and Fritz London introduce the concepts of valence bond theory and apply it to the hydrogen molecule.

- 1927 – Thomas and Fermi develop the Thomas–Fermi model for a Gas in a box.

- 1927 – Chandrasekhara Venkata Raman studies optical photon scattering by electrons.

- 1927 – Dirac states his relativistic electron quantum wave equation, the Dirac equation.

- 1927 – Charles G. Darwin and Walter Gordon solve the Dirac equation for a Coulomb potential.

- 1927 – Charles Drummond Ellis (along with James Chadwick and colleagues) finally establish clearly that the beta decay spectrum is in fact continuous and not discrete, posing a problem that will later be solved by theorizing (and later discovering) the existence of the neutrino.

- 1927 – Walter Heitler uses Schrödinger's wave equation to show how two hydrogen atom wavefunctions join together, with plus, minus, and exchange terms, to form a covalent bond.

- 1927 – Robert Mulliken works, in coordination with Hund, to develop a molecular orbital theory where electrons are assigned to states that extend over an entire molecule and, in 1932, introduces many new molecular orbital terminologies, such as σ bond, π bond, and δ bond.

- 1927 – Eugene Wigner relates degeneracies of quantum states to irreducible representations of symmetry groups.

- 1927 – Hermann Klaus Hugo Weyl proves in collaboration with his student Fritz Peter a fundamental theorem in harmonic analysis—the Peter–Weyl theorem—relevant to group representations in quantum theory (including the complete reducibility of unitary representations of a compact topological group);[17] introduces the Weyl quantization, and earlier, in 1918, introduces the concept of gauge and a gauge theory; later in 1935 he introduces and characterizes with Richard Bauer the concept of spinor in n-dimensions.[18]

- 1928 – Linus Pauling outlines the nature of the chemical bond: uses Heitler's quantum mechanical covalent bond model to outline the quantum mechanical basis for all types of molecular structure and bonding and suggests that different types of bonds in molecules can become equalized by rapid shifting of electrons, a process called "resonance" (1931), such that resonance hybrids contain contributions from the different possible electronic configurations.

- 1928 – Friedrich Hund and Robert S. Mulliken introduce the concept of molecular orbitals.

- 1928 – Born and Vladimir Fock formulate and prove the adiabatic theorem, which states that a physical system shall remain in its instantaneous eigenstate if a given perturbation is acting on it slowly enough and if there is a gap between the eigenvalue and the rest of the Hamiltonian's spectrum.

- 1929 – Oskar Klein discovers the Klein paradox

- 1929 – Oskar Klein and Yoshio Nishina derive the Klein–Nishina cross section for high energy photon scattering by electrons

- 1929 – Sir Nevill Mott derives the Mott cross section for the Coulomb scattering of relativistic electrons

- 1929 – John Lennard-Jones introduces the linear combination of atomic orbitals approximation for the calculation of molecular orbitals.

- 1929 – Fritz Houtermans and Robert d'Escourt Atkinson propose that stars release energy by nuclear fusion.[1]

16.2.4 1930–1939

- 1930 – Dirac hypothesizes the existence of the positron.[1]

- 1930 – Dirac's textbook *Principles of Quantum Mechanics* is published, becoming a standard reference book that is still used today.

- 1930 – Erich Hückel introduces the Hückel molecular orbital method, which expands on orbital theory to determine the energies of orbitals of pi electrons in conjugated hydrocarbon systems.

- 1930 – Fritz London explains van der Waals forces as due to the interacting fluctuating dipole moments between molecules

- 1930 – Pauli suggests in a famous letter that, in addition to electrons and protons, atoms also contain an extremely light neutral particle which he calls the "neutron." He suggests that this "neutron" is also emitted during beta decay and has simply not yet been observed. Later it is determined that this particle is actually the almost massless neutrino.[1]

- 1931 – John Lennard-Jones proposes the Lennard-Jones interatomic potential

- 1931 – Walther Bothe and Herbert Becker find that if the very energetic alpha particles emitted from polonium fall on certain light elements, specifically beryllium, boron, or lithium, an unusually penetrating radiation is produced. At first this radiation is thought to be gamma radiation, although it is more penetrating than any gamma rays known, and the details of experimental results are very difficult to interpret on this basis. Some scientists begin to hypothesize the possible existence of another fundamental particle.

- 1931 – Erich Hückel redefines the property of aromaticity in a quantum mechanical context by introducing the 4n+2 rule, or Hückel's rule, which predicts whether an organic planar ring molecule will have aromatic properties.

- 1931 – Ernst Ruska creates the first electron microscope.[1]

- 1931 – Ernest Lawrence creates the first cyclotron and founds the Radiation Laboratory, later the Lawrence Berkeley National Laboratory; in 1939 he awarded the Nobel Prize in Physics for his work on the cyclotron.

- 1932 – Irène Joliot-Curie and Frédéric Joliot show that if the unknown radiation generated by alpha particles falls on paraffin or any other hydrogen-containing compound, it ejects protons of very high energy. This is not in itself inconsistent with the proposed gamma ray nature of the new radiation, but detailed quantitative analysis of the data become increasingly difficult to reconcile with such a hypothesis.

- 1932 – James Chadwick performs a series of experiments showing that the gamma ray hypothesis for the unknown radiation produced by alpha particles is untenable, and that the new particles must be the neutrons hypothesized by Fermi.[1]

- 1932 – Werner Heisenberg applies perturbation theory to the two-electron problem to show how resonance arising from electron exchange can explain exchange forces.

- 1932 – Mark Oliphant: Building upon the nuclear transmutation experiments of Ernest Rutherford done a few years earlier, observes fusion of light nuclei (hydrogen isotopes). The steps of the main cycle of nuclear fusion in stars are subsequently worked out by Hans Bethe over the next decade.

- 1932 – Carl D. Anderson experimentally proves the existence of the positron.[1]

- 1933 – Following Chadwick's experiments, Fermi renames Pauli's "neutron" to neutrino to distinguish it from Chadwick's theory of the much more massive neutron.

- 1933 – Leó Szilárd first theorizes the concept of a nuclear chain reaction. He files a patent for his idea of a simple nuclear reactor the following year.

- 1934 – Fermi publishes a very successful model of beta decay in which neutrinos are produced.

- 1934 – Fermi studies the effects of bombarding uranium isotopes with neutrons.

- 1934 – N. N. Semyonov develops the total quantitative chain chemical reaction theory, later the basis of various high technologies using the incineration of gas mixtures. The idea is also used for the description of the nuclear reaction.

- 1934 – Irène Joliot-Curie and Frédéric Joliot-Curie discover artificial radioactivity and are jointly awarded the 1935 Nobel Prize in Chemistry[19]

- 1935 – Einstein, Boris Podolsky, and Nathan Rosen describe the EPR paradox which challenges the completeness of quantum mechanics as it was theorized up to that time. Assuming that local realism is valid, they demonstrated that there would need to be hidden parameters to explain how measuring the quantum state of one particle could influence the quantum state of another particle without apparent contact between them.[20]

- 1935 - Schrödinger develops the Schrödinger's cat thought experiment. It illustrates what he saw as the problems of the Copenhagen interpretation of quantum mechanics if subatomic particles can be in two contradictory quantum states at once.

- 1935 – Hideki Yukawa formulates his hypothesis of the Yukawa potential and predicts the existence of the pion, stating that such a potential arises from the exchange of a massive scalar field, as it would be found in the field of the pion. Prior to Yukawa's paper, it was believed that the scalar fields of the fundamental forces necessitated massless particles.

- 1936 – Alexandru Proca publishes prior to Hideki Yukawa his relativistic quantum field equations for a massive vector meson of spin-1 as a basis for nuclear forces.

- 1936 – Garrett Birkhoff and John von Neumann introduce Quantum Logic[21] in an attempt to reconcile the apparent inconsistency of classical, Boolean logic with the Heisenberg Uncertainty Principle of quantum mechanics as applied, for example, to the measurement of complementary (noncommuting) observables in quantum mechanics, such as position and momentum;[22] current approaches to quantum logic involve noncommutative and non-associative many-valued logic.[23][24]

- 1936 – Carl D. Anderson discovers muons while he is studying cosmic radiation.

- 1937 – Carl Anderson experimentally proves the existence of the pion.

- 1937 – Hermann Arthur Jahn and Edward Teller prove, using group theory, that non-linear degenerate molecules are unstable.[25] The Jahn-Teller theorem essentially states that any non-linear molecule with a degenerate electronic ground state will undergo a geometrical distortion that removes that degeneracy, because the distortion lowers the overall energy of the complex. The latter process is called the Jahn-Teller effect; this effect was recently considered also in relation to the superconductivity mechanism in YBCO and other high temperature superconductors. The details of the Jahn-Teller effect are presented with several examples and EPR data in the basic textbook by Abragam and Bleaney (1970).

- 1938 – Charles Coulson makes the first accurate calculation of a molecular orbital wavefunction with the hydrogen molecule.

- 1938 – Otto Hahn and his assistant Fritz Strassmann send a manuscript to Naturwissenschaften reporting they have detected the element barium after bombarding uranium with neutrons. Hahn calls this new phenomenon a 'bursting' of the uranium nucleus. Simultaneously, Hahn communicate these results to Lise Meitner. Meitner, and her nephew Otto Robert Frisch, correctly interpret these results as being a nuclear fission. Frisch confirms this experimentally on 13 January 1939.

- 1939 – Leó Szilárd and Fermi discover neutron multiplication in uranium, proving that a chain reaction is indeed possible.

16.2.5 1940–1949

- 1942 – Kan-Chang Wang first proposes the use of K-electron capture to experimentally detect neutrinos.

- 1942 – A team led by Enrico Fermi creates the first artificial self-sustaining nuclear chain reaction, called Chicago Pile-1, in a racquets court below the bleachers of Stagg Field at the University of Chicago on December 2, 1942.

- 1942 to 1946 – J. Robert Oppenheimer successfully leads the Manhattan Project, predicts quantum tunneling and proposes the Oppenheimer–Phillips process in nuclear fusion

- 1945 – the Manhattan Project produces the first nuclear fission explosion on July 16, 1945 in the Trinity test in New Mexico.

- 1945 – John Archibald Wheeler and Richard Feynman originate Wheeler–Feynman absorber theory, an interpretation of electrodynamics that supposes that elementary particles are not self-interacting.

- 1946 – Theodor V. Ionescu and Vasile Mihu report the construction of the first hydrogen maser by stimulated emission of radiation in molecular hydrogen.

- 1947 – Willis Lamb and Robert Retherford measure a small difference in energy between the energy levels $^2S_1/2$ and $^2P_1/2$ of the hydrogen atom, known as the Lamb shift.

- 1947 – George Rochester and Clifford Charles Butler publishes two cloud chamber photographs of cosmic ray-induced events, one showing what appears to be a neutral particle decaying into two charged pions, and one that appears to be a charged particle decaying into a charged pion and something neutral. The estimated mass of the new particles is very rough, about half a proton's mass. More examples of these "V-particles" were slow in coming, and they are soon given the name kaons.

- 1948 – Sin-Itiro Tomonaga and Julian Schwinger Independently introduce perturbative renormalization as a method of correcting the original Lagrangian of a quantum field theory so as to eliminate a series of infinite terms that would otherwise result.

- 1948 – Richard Feynman states the path integral formulation of quantum mechanics.

- 1949 – Freeman Dyson determines the equivalence of two formulations of quantum electrodynamics: Feynman's diagrammatic path integral formulation and the operator method developed by Julian Schwinger and Tomonaga. A by-product of that demonstration is the invention of the Dyson series.[26]

16.2.6 1950–1959

- 1951 – Clemens C. J. Roothaan and George G. Hall derive the Roothaan-Hall equations, putting rigorous molecular orbital methods on a firm basis.

- 1951 – Edward Teller, physicist and "father of the hydrogen bomb", and Stanislaw Ulam, mathematician, are reported to have written jointly in March 1951 a classified report on "Hydrodynamic Lenses and Radiation Mirrors" that results in the next step in the Manhattan Project.[27]

- 1951 and 1952 – at the Manhattan Project, the first planned fusion thermonuclear reaction experiment is carried out successfully in the Spring of 1951 at Eniwetok, based on the work of Edward Teller and Dr. Hans A. Bethe.[28] The Los Alamos Laboratory proposes a date in November 1952 for a hydrogen bomb, full-scale test that is apparently carried out.

- 1951 – Felix Bloch and Edward Mills Purcell receive a shared Nobel Prize in Physics for their first observations of the quantum phenomenon of nuclear magnetic resonance previously reported in 1949.[29][30][31] Purcell reports his contribution as *Research in Nuclear Magnetism*, and gives credit to his coworkers such as Herbert S. Gutowsky for their NMR contributions,[32][33] as well as theoretical researchers of nuclear magnetism such as John Hasbrouck Van Vleck.

- 1952 – Albert W. Overhauser formulates a theory of dynamic nuclear polarization, also known as the Overhauser Effect; other contenders are the subsequent theory of Ionel Solomon reported in 1955 that includes the *Solomon equations* for the dynamics of coupled spins, and that of R. Kaiser in 1963. The general Overhauser effect is first demonstrated experimentally by T. R. Carver and Charles P. Slichter in 1953.[34]

- 1952 – Donald A. Glaser creates the bubble chamber, which allows detection of electrically charged particles by surrounding them by a bubble. Properties of the particles such as momentum can be determined by studying of their helical paths. Glaser receives a Nobel prize in 1960 for his invention.

- 1953 – Charles H. Townes, collaborating with James P. Gordon, and H. J. Zeiger, builds the first ammonia maser; receives a Nobel prize in 1964 for his experimental success in producing coherent radiation by atoms and molecules.

- 1954 – Chen Ning Yang and Robert Mills derive a gauge theory for nonabelian groups, leading to the successful formulation of both electroweak unification and quantum chromodynamics.

- 1955 – Ionel Solomon develops the first nuclear magnetic resonance theory of magnetic dipole coupled nuclear spins and of the Nuclear Overhauser Effect.

- 1955 and 1956 – Murray Gell-Mann and Kazuhiko Nishijima independently derive the Gell-Mann–Nishijima formula, which relates the baryon number, the strangeness, and the isospin of hadrons to the charge, eventually leading to the systematic categorization of hadrons and, ultimately, the Quark Model of hadron composition.

- 1956 – P. Kuroda predicts that self-sustaining nuclear chain reactions should occur in natural uranium deposits.

- 1956 – Chien-Shiung Wu carries out the Wu Experiment, which observes parity violation in cobalt-60 decay, showing that parity violation is present in the weak interaction.

- 1956 – Clyde L. Cowan and Frederick Reines experimentally prove the existence of the neutrino.

- 1957 – John Bardeen, Leon Cooper and John Robert Schrieffer propose their quantum BCS theory of low temperature superconductivity, for which their receive a Nobel prize in 1972. The theory represents superconductivity as a macroscopic quantum coherence phenomenon involving phonon coupled electron pairs with opposite spin

- 1957 – William Alfred Fowler, Margaret Burbidge, Geoffrey Burbidge, and Fred Hoyle, in their 1957 paper *Synthesis of the Elements in Stars*, show that the abundances of essentially all but the lightest chemical elements can be explained by the process of nucleosynthesis in stars.

- 1957 – Hugh Everett formulates the many-worlds interpretation of quantum mechanics, which states that every possible quantum outcome is realized in divergent, non-communicating parallel universes in quantum superposition.[35][36]

- 1958–1959 – magic angle spinning described by Edward Raymond Andrew, A. Bradbury, and R. G. Eades, and independently in 1959 by I. J. Lowe.[37]

16.2.7   1960–1969

- 1961 – Clauss Jönsson performs Young's double-slit experiment (1909) for the first time with particles other than photons by using electrons and with similar results, confirming that massive particles also behaved according to the wave–particle duality that is a fundamental principle of quantum field theory.

- 1961 – Anatole Abragam publishes the fundamental textbook on the quantum theory of Nuclear Magnetic Resonance entitled *The Principles of Nuclear Magnetism*;[39]

- 1961 – Sheldon Lee Glashow extends the electroweak interaction modelss developed by Julian Schwinger by including a short range neutral current, the Z_o. The resulting symmetry structure that Glashow proposes, SU(2) X U(1), forms the basis of the accepted theory of the electroweak interactions.

- 1962 – Leon M. Lederman, Melvin Schwartz and Jack Steinberger show that more than one type of neutrino exists by detecting interactions of the muon neutrino (already hypothesised with the name "neutretto")

- 1962 – Murray Gell-Mann and Yuval Ne'eman independently classify the hadrons according to a system that Gell-Mann called the Eightfold Way, and which ultimately led to the quark model (1964) of hadron composition.

- 1962 – Jeffrey Goldstone, Yoichiro Nambu, Abdus Salam, and Steven Weinberg develop what is now known as Goldstone's Theorem: if there is a continuous symmetry transformation under which the Lagrangian is invariant, then either the vacuum state is also invariant under the transformation, or there must be spinless particles of zero mass, thereafter called Nambu-Goldstone bosons.

- 1962 to 1973 – Brian David Josephson, predicts correctly the quantum tunneling effect involving superconducting currents while he is a PhD student under the supervision of Professor Brian Pippard at the Royal Society Mond Laboratory in Cambridge, UK; subsequently, in 1964, he applies his theory to coupled superconductors. The effect is later demonstrated experimentally at Bell Labs in the USA. For his important quantum discovery he is awarded the Nobel Prize in Physics in 1973.[40]

- 1963 – Eugene P. Wigner lays the foundation for the theory of symmetries in quantum mechanics as well as for basic research into the structure of the atomic nucleus; makes important "contributions to the theory of the atomic nucleus and the elementary particles, particularly through the discovery and application of fundamental symmetry principles"; he shares half of his Nobel prize in Physics with Maria Goeppert-Mayer and J. Hans D. Jensen.

- 1963 – Maria Goeppert Mayer and J. Hans D. Jensen share with Eugene P. Wigner half of the Nobel Prize in Physics in 1963 "for their discoveries concerning nuclear shell structure theory".[41]

- 1963 – Nicola Cabibbo develops the mathematical matrix by which the first two (and ultimately three) generations of quarks can be predicted.

- 1964 – Murray Gell-Mann and George Zweig independently propose the quark model of hadrons, predicting the arbitrarily named up, down, and strange quarks. Gell-Mann is credited with coining the term *quark*, which he found in James Joyce's book *Finnegans Wake*.

- 1964 – François Englert, Robert Brout, Peter Higgs, Gerald Guralnik, C. R. Hagen, and Tom Kibble postulate that a fundamental quantum field, now called the Higgs field, permeates space and, by way of the Higgs mechanism, provides mass to all the elementary subatomic particles that interact with it. While the Higgs field is postulated to confer mass on quarks and leptons, it represents only a tiny portion of the masses of other subatomic particles, such as protons and neutrons. In these, gluons that bind quarks together confer most of the particle mass. The result is obtained independently by three groups: François Englert and Robert Brout; Peter Higgs, working from the ideas of Philip Anderson; and Gerald Guralnik, C. R. Hagen, and Tom Kibble.[42][43][44][45][46][47][48]

- 1964 – Sheldon Lee Glashow and James Bjorken predict the existence of the charm quark. The addition is proposed because it allows for a better description of the weak interaction (the mechanism that allows quarks and other particles to decay), equalizes the number of known quarks with the number of known leptons, and implies a mass formula that correctly reproduced the masses of the known mesons.

- 1964 – John Stewart Bell puts forth Bell's theorem, which used testable inequality relations to show the flaws in the earlier Einstein–Podolsky–Rosen paradox and prove that no physical theory of local hidden variables can ever reproduce all of the predictions of quantum mechanics. This inaugurated the study of quantum entanglement, the phenomenon in which separate particles share the same quantum state despite being at a distance from each other.

- 1964 – Nikolai G. Basov and Aleksandr M. Prokhorov share the Nobel Prize in Physics in 1964 for, respectively, semiconductor lasers and Quantum Electronics; they also share the prize with Charles Hard Townes, the inventor of the ammonium maser.

- 1967 – Steven Weinberg and Abdus Salam publish a paper in which he describes Yang–Mills theory using the SU(2) X U(1) supersymmetry group, thereby yielding a mass for the W particle of the weak interaction via spontaneous symmetry breaking.

- 1968 – Stanford University: Deep inelastic scattering experiments at the Stanford Linear Accelerator Center (SLAC) show that the proton contains much smaller, point-like objects and is therefore not an elementary particle. Physicists at the time are reluctant to identify these objects with quarks, instead calling them *partons* —

a term coined by Richard Feynman. The objects that are observed at SLAC will later be identified as up and down quarks. Nevertheless, "parton" remains in use as a collective term for the constituents of hadrons (quarks, antiquarks, and gluons). The existence of the strange quark is indirectly validated by the SLAC's scattering experiments: not only is it a necessary component of Gell-Mann and Zweig's three-quark model, but it provides an explanation for the kaon (K) and pion (π) hadrons discovered in cosmic rays in 1947.

- 1969 to 1977 – Sir Nevill Mott and Philip Warren Anderson publish quantum theories for electrons in non-crystalline solids, such as glasses and amorphous semiconductors; receive in 1977 a Nobel prize in Physics for their investigations into the electronic structure of magnetic and disordered systems, which allow for the development of electronic switching and memory devices in computers. The prize is shared with John Hasbrouck Van Vleck for his contributions to the understanding of the behavior of electrons in magnetic solids; he established the fundamentals of the quantum mechanical theory of magnetism and the crystal field theory (chemical bonding in metal complexes) and is regarded as the Father of modern Magnetism.

- 1969 and 1970 – Theodor V. Ionescu, Radu Pârvan and I.C. Baianu observe and report quantum amplified stimulation of electromagnetic radiation in hot deuterium plasmas in a longitudinal magnetic field; publish a quantum theory of the amplified coherent emission of radiowaves and microwaves by focused electron beams coupled to ions in hot plasmas.

- 1970 – Glashow, John Iliopoulos and Luciano Maiani predict the charmed quark that is subsequently found experimentally and share a Nobel prize for their theoretical prediction.

16.2.8   1971–1979

- 1971 – Martinus J. G. Veltman and Gerardus 't Hooft show that, if the symmetries of Yang–Mills theory are broken according to the method suggested by Peter Higgs, then Yang–Mills theory can be renormalized. The renormalization of Yang–Mills Theory predicts the existence of a massless particle, called the gluon, which could explain the nuclear strong force. It also explains how the particles of the weak interaction, the W and Z bosons, obtain their mass via spontaneous symmetry breaking and the Yukawa interaction.

- 1972 – Francis Perrin discovers "natural nuclear fission reactors" in uranium deposits in Oklo, Gabon, where analysis of isotope ratios demonstrate that self-sustaining, nuclear chain reactions have occurred. The conditions under which a natural nuclear reactor could exist were predicted in 1956 by P. Kuroda.

- 1973 – Frank Anthony Wilczek discover the quark asymptotic freedom in the theory of strong interactions; receives the Lorentz Medal in 2002, and the Nobel Prize in Physics in 2004 for his discovery and his subsequent contributions to quantum chromodynamics.[50]

- 1973 – Makoto Kobayashi and Toshihide Maskawa note that the experimental observation of CP violation can be explained if an additional pair of quarks exist. The two new quarks are eventually named top and bottom.

- 1973 – Peter Mansfield formulates the physical theory of Nuclear magnetic resonance imaging (NMRI)[51][52][53][54]

- 1974 – Pier Giorgio Merli performs Young's double-slit experiment (1909) using a single electron with similar results, confirming the existence of quantum fields for massive particles.

- 1974 – Burton Richter and Samuel Ting: Charm quarks are produced almost simultaneously by two teams in November 1974 (see November Revolution) — one at SLAC under Burton Richter, and one at Brookhaven National Laboratory under Samuel Ting. The charm quarks are observed bound with charm antiquarks in mesons. The two discovering parties independently assign the discovered meson two different symbols, J and ψ; thus, it becomes formally known as the J/ψ meson. The discovery finally convinces the physics community of the quark model's validity.

- 1975 – Martin Lewis Perl, with his colleagues at the SLAC–LBL group, detects the tau in a series of experiments between 1974 and 1977.

- 1977 – Leon Lederman observes the bottom quark with his team at Fermilab. This discovery is a strong indicator of the top quark's existence: without the top quark, the bottom quark would be without a partner that is required by the mathematics of the theory.

- 1977 – Ilya Prigogine develops non-equilibrium, irreversible thermodynamics and quantum operator theory, especially the time superoperator theory; he is awarded the Nobel Prize in Chemistry in 1977 "for his contributions to non-equilibrium thermodynamics, particularly the theory of dissipative structures".[55]

- 1978 – Pyotr Kapitsa observes new phenomena in hot deuterium plasmas excited by very high power microwaves in attempts to obtain controlled thermonuclear fusion reactions in such plasmas placed in longitudinal magnetic fields, using a novel and low-cost design of thermonuclear reactor, similar in concept to that reported by Theodor V. Ionescu *et al.* in 1969. Receives a Nobel prize for early low temperature physics experiments on helium superfluidity carried out in 1937 at the Cavendish Laboratory in Cambridge, UK, and discusses his 1977 thermonuclear reactor results in his Nobel lecture on December 8, 1978.

- 1979 – Kenneth A. Rubinson and coworkers, at the Cavendish Laboratory, observe ferromagnetic spin wave resonant excite journals (FSWR) in locally anisotropic, FENiPB metallic glasses and interpret the observations in terms of two-magnon dispersion and a spin exchange Hamiltonian, similar in form to that of a Heisenberg ferromagnet.[56]

16.2.9   1980–1999

- 1980 to 1982 – Alain Aspect verify experimentally the quantum entanglement hypothesis; his Bell test experiments provide strong evidence that a quantum event at one location can affect an event at another location without any obvious mechanism for communication between the two locations.[57][58]

- 1982 to 1997 – Tokamak Fusion Test Reactor (TFTR) at PPPL, Princeton, USA: Operated since 1982, produces 10.7MW of controlled fusion power for only 0.21s in 1994 by using T-D nuclear fusion in a tokamak reactor with "a toroidal 6T magnetic field for plasma confinement, a 3MA plasma current and an electron density of 1.0×10^{20} m^{-3} of 13.5 keV" [59]

- 1983 – Carlo Rubbia and Simon van der Meer, at the Super Proton Synchrotron, see unambiguous signals of W particles in January. The actual experiments are called UA1 (led by Rubbia) and UA2 (led by Peter Jenni), and are the collaborative effort of many people. Simon van der Meer is the driving force on the use of the accelerator. UA1 and UA2 find the Z particle a few months later, in May 1983.

- 1983 to 2011 – The largest and most powerful experimental nuclear fusion tokamak reactor in the world, Joint European Torus (JET) begins operation at Culham Facility in UK; operates with T-D plasma pulses and has a reported gain factor Q of 0.7 in 2009, with an input of 40MW for plasma heating, and a 2800-ton iron magnet for confinement;[60] in 1997 in a tritium-deuterium experiment JET produces 16 MW of fusion power, a total of 22 MJ of fusion, energy and a steady fusion power of 4 MW which is maintained for 4 seconds.[61]

- 1985 to 2010 – The JT-60 (Japan Torus) begins operation in 1985 with an experimental D-D nuclear fusion tokamak similar to the JET; in 2010 JT-60 holds the record for the highest value of the fusion triple product achieved: 1.77×10^{28} K·s·m^{-3} = 1.53×10^{21} keV·s·m^{-3}.;[62] JT-60 claims it would have an equivalent energy gain factor, Q of 1.25 if it were operated with a T-D plasma instead of the D-D plasma, and on May 9, 2006 attains a fusion hold time of 28.6 s in full operation; moreover, a high-power microwave gyrotron construction is completed that is capable of *1.5MW* output for *1s*,[63] thus meeting the conditions for the planned ITER, large-scale nuclear fusion reactor. JT-60 is disassembled in 2010 to be upgraded to a more powerful nuclear fusion reactor—the JT-60SA—by using niobium-titanium superconducting coils for the magnet confining the ultra-hot D-D plasma.

- 1986 – Johannes Georg Bednorz and Karl Alexander Müller produce unambiguous experimental proof of high temperature superconductivity involving Jahn-Teller polarons in orthorhombic La_2CuO_4, YBCO and other perovskite-type oxides; promptly receive a Nobel prize in 1987 and deliver their Nobel lecture on December 8, 1987.[64]

- 1986 – Vladimir Gershonovich Drinfeld introduces the concept of quantum groups as Hopf algebras in his seminal address on quantum theory at the International Congress of Mathematicians, and also connects them to the study of the Yang–Baxter equation, which is a necessary condition for the solvability of statistical mechanics models; he also generalizes Hopf algebras to quasi-Hopf algebras, and introduces the study of Drinfeld twists, which can be used to factorize the R-matrix corresponding to the solution of the Yang–Baxter equation associated with a quasitriangular Hopf algebra.

- 1988 to 1998 – Mihai Gavrilă discovers in 1988 the new quantum phenomenon of *atomic dichotomy* in hydrogen and subsequently publishes a book on the atomic structure and decay in high-frequency fields of hydrogen atoms placed in ultra-intense laser fields.[65][66][67][68][69][70][71]

- 1991 – Richard R. Ernst develops two-dimensional nuclear magnetic resonance spectroscopy (2D-FT NMRS) for small molecules in solution and is awarded the Nobel Prize in Chemistry in 1991 "for his contributions to the development of the methodology of high resolution nuclear magnetic resonance (NMR) spectroscopy."[72]

- 1977 to 1995 – The top quark is finally observed by a team at Fermilab after an 18-year search. It has a mass much greater than had been previously expected — almost as great as a gold atom.

- 1995 – Eric Cornell, Carl Wieman and Wolfgang Ketterle and co-workers at JILA create the first "pure" Bose–Einstein condensate. They do this by cooling a dilute vapor consisting of approximately two thousand rubidium-87 atoms to below 170 nK using a combination of laser cooling and magnetic evaporative cooling. About four months later, an independent effort led by Wolfgang Ketterle at MIT creates a condensate made of sodium-23. Ketterle's condensate has about a hundred times more atoms, allowing him to obtain several important results such as the observation of quantum mechanical interference between two different condensates.

- 1998 – The Super-Kamiokande (Japan) detector facility reports experimental evidence for neutrino oscillations, implying that at least one neutrino has mass.

- 1999 to 2013 – NSTX—The National Spherical Torus Experiment at PPPL, Princeton, USA launches a nuclear fusion project on February 12, 1999 for "an innovative magnetic fusion device that was constructed by the Princeton Plasma Physics Laboratory (PPPL) in collaboration with the Oak Ridge National Laboratory, Columbia University, and the University of Washington at Seattle"; NSTX is being used to study the physics principles of spherically shaped plasmas.[73]

16.3 21st century

- 2000 – scientists at European Organization for Nuclear Research (CERN) publish experimental results in which they claim to have observed indirect evidence of the existence of a quark–gluon plasma, which they call a "new state of matter."

- 2001 – the Sudbury Neutrino Observatory (Canada) confirm the existence of neutrino oscillations. Lene Hau stops a beam of light completely in a Bose–Einstein condensate.[74]

- 2002 – Leonid Vainerman organizes a meeting at Strasbourg of theoretical physicists and mathematicians focused on quantum group and quantum groupoid applications in quantum theories; the proceedings of the meeting are published in 2003 in a book edited by the meeting organizer.[75]

- 2003 – Sir Anthony James Leggett receives the 2003 Nobel Prize in Physics for pioneering contributions to the quantum theory of superconductors, and superfluids such as Helium-3, shared with V. L. Ginzburg and A. A. Abrikosov.

- 2005 – the RHIC accelerator of Brookhaven National Laboratory generates a quark-gluon fluid, perhaps the quark–gluon plasma

- 2007 to 2010 – Charles Pence Slichter is awarded the National Medal of Science in 2007 for his studies of Nuclear Magnetic Resonance in Solids, and especially his NMR Studies of High-Temperature Superconductors.

- 2008 to 2010 – the Lithium Tokamak Experiment (LTX) starts in September 2008.[76]

- 2007 to 2010 – Alain Aspect, Anton Zeilinger and John Clauser present progress with the resolution of the non-locality aspect of quantum theory and in 2010 are awarded the Wolf Prize in Physics, together with Anton Zeilinger and John Clauser.[77]

- 2009 - Aaron D. O'Connell invents the first quantum machine, applying quantum mechanics to a macroscopic object just large enough to be seen by the naked eye, which is able to vibrate a small amount and large amount simultaneously.

- 2010 – Andre Geim and Konstantin Novoselov receive the Nobel Prize in Physics "for groundbreaking experiments regarding the two-dimensional material graphene".

- 2011 - Zachary Dutton demonstrates how photons can co-exist in superconductors. "Direct Observation of Coherent Population Trapping in a Superconducting Artificial Atom",[78]

- 2014 – Scientists transfer data by quantum teleportation over a distance of 10 feet with zero percent error rate, a vital step towards a quantum internet.[79][80]

16.4 See also

- History of quantum mechanics
- Timeline of atomic and subatomic physics
- Timeline of particle physics
- Timeline of physical chemistry

Learning materials related to the history of Quantum Mechanics at Wikiversity

16.5 References

[1] Peacock 2008, pp. 175–183

[2] Ben-Menahem 2009

[3] Becquerel, Henri (1896). "Sur les radiations émises par phosphorescence". *Comptes Rendus* **122**: 420–421.

[4] Marie Curie and the Science of Radioactivity: Research Breakthroughs (1897–1904). Aip.org. Retrieved on 2012-05-17.

[5] Frederick Soddy (December 12, 1922). "The origins of the conceptions of isotopes" (PDF). *Nobel Lecture in Chemistry*. Retrieved April 2012.

[6] Ernest Rutherford, Baron Rutherford of Nelson, of Cambridge. Encyclopædia Britannica on-line. Retrieved on 2012-05-17.

[7] The Nobel Prize in Chemistry 1908: Ernest Rutherford. nobelprize.org

[8] McCormmach, Russell (Spring 1967). "Henri Poincaré and the Quantum Theory". *Isis* **58** (1): 37–55. doi:10.1086/350182.

[9] Irons, F. E. (August 2001). "Poincaré's 1911–12 proof of quantum discontinuity interpreted as applying to atoms". *American Journal of Physics* **69** (8): 879–884. Bibcode:2001AmJPh..69..879I. doi:10.1119/1.1356056.

[10] Ştefan Procopiu. 1913. "Determining the Molecular Magnetic Moment by M. Planck's Quantum Theory". *Bulletin scientifique de l'Académie Roumaine de sciences.*, 1:151.

[11] Pais, Abraham (1995). "Introducing Atoms and Their Nuclei". In Brown, Laurie M.; Pais, Abraham; Pippard, Brian. *Twentieth Century Physics* **1**. American Institute of Physics Press. p. 89. ISBN 9780750303101. Now the beauty of Franck and Hertz's work lies not only in the measurement of the energy loss E_2-E_1 of the impinging electron, but they also observed that, when the energy of that electron exceeds 4.9 eV, mercury begins to emit ultraviolet light of a definite frequency ν as defined in the above formula. Thereby they gave (unwittingly at first) the first direct experimental proof of the Bohr relation!

[12] P. S. Epstein, *Zur Theorie des Starkeffektes*, Annalen der Physik, vol. **50**, pp. 489-520 (1916)

[13] K. Schwarzschild, Sitzungsberichten der Kgl. Preuss. Akad. d. Wiss. April 1916, p. 548

[14] Lewis, G.N. (1926)."The conservation of photons".*Nature***118**(2981): 874–875.Bibcode:1926Natur.118..874L.doi:10.1

[15] P. S. Epstein, *The Stark Effect from the Point of View of Schroedinger's Quantum Theory*, Physical Review, vol **28**, pp. 695-710 (1926)

[16] John von Neumann. 1932. *The Mathematical Foundations of Quantum Mechanics.*, Princeton University Press: Princeton, New Jersey, reprinted in 1955, 1971 and 1983 editions

[17] Peter, F.; Weyl, H. (1927). "Die Vollständigkeit der primitiven Darstellungen einer geschlossenen kontinuierlichen Gruppe". *Math. Ann.* **97**: 737–755. doi:10.1007/BF01447892.

[18] Brauer, Richard; Weyl, Hermann (1935). "Spinors in n dimensions". *American Journal of Mathematics* (The Johns Hopkins University Press) **57** (2): 425–449. doi:10.2307/2371218. JSTOR 2371218.

[19] Frédéric Joliot-Curie (December 12, 1935). "Chemical evidence of the transmutation of elements" (PDF). *Nobel Lecture*. Retrieved April 2012.

[20] Einstein A, Podolsky B, Rosen N; Podolsky; Rosen (1935). "Can Quantum-Mechanical Description of Physical Reality Be Considered Complete?". *Phys. Rev.* **47** (10): 777–780. Bibcode:1935PhRv...47..777E. doi:10.1103/PhysRev.47.777.

[21] Birkhoff, Garrett and von Neumann, J. (1936). "The Logic of Quantum Mechanics". *Annals of Mathematics* **37** (4): 823–843. doi:10.2307/1968621. JSTOR 1968621.

[22] Roland Omnès (8 March 1999). *Understanding Quantum Mechanics*. Princeton University Press. ISBN 978-0-691-00435-8. Retrieved 17 May 2012.

[23] Dalla Chiara, M. L.; Giuntini, R. (1994). "Unsharp quantum logics".*Foundations of Physics***24**(8): 1161–1177.Bibcode:1994 doi:10.1007/BF02057862.

[24] Georgescu, G. (2006). "N-valued Logics and Łukasiewicz-Moisil Algebras". *Axiomathes* **16** (1–2): 123. doi:10.1007/s10516-005-4145-6.

[25] H. Jahn and E. Teller (1937). "Stability of Polyatomic Molecules in Degenerate Electronic States. I. Orbital Degeneracy". *Proceedings of the Royal Society A* **161** (905): 220–235. Bibcode:1937RSPSA.161..220J. doi:10.1098/rspa.1937.0142.

[26] Dyson, F. (1949). "The S Matrix in Quantum Electrodynamics". *Phys. Rev.* **75** (11): 1736. Bibcode:1949PhRv...75.1736D. doi:10.1103/PhysRev.75.1736.

[27] Stix, Gary (October 1999). "Infamy and honor at the Atomic Café: Edward Teller has no regrets about his contentious career". *Scientific American*: 42–43. Retrieved April 2012.

[28] Hans A. Bethe (May 28, 1952). MEMORANDUM ON THE HISTORY OF THERMONUCLEAR PROGRAM (Report). Reconstructed version from only partially declassified documents, with certain words deliberately deleted.

[29] Bloch, F.; Hansen, W.; Packard, Martin (1946). "Nuclear Induction".*Physical Review***69**(3–4): 127.Bibcode:1946PhRv...69 doi:10.1103/PhysRev.69.127.

[30] Bloch, F.; Jeffries, C. (1950). "A Direct Determination of the Magnetic Moment of the Proton in Nuclear Magnetons". *Physical Review* **80** (2): 305. Bibcode:1950PhRv...80..305B. doi:10.1103/PhysRev.80.305.

[31] Bloch, F. (1946). "Nuclear Induction".*Physical Review***70**(7–8): 460.Bibcode:1946PhRv...70..460B.doi:10.1103/PhysRev.

[32] Gutowsky, H. S.; Kistiakowsky, G. B.; Pake, G. E.; Purcell, E. M. (1949). "Structural Investigations by Means of Nuclear Magnetism. I. Rigid Crystal Lattices". *The Journal of Chemical Physics* **17** (10): 972. Bibcode:1949JChPh..17..972G. doi:10.1063/1.1747097.

[33] Gardner, J.; Purcell, E. (1949). "A Precise Determination of the Proton Magnetic Moment in Bohr Magnetons". *Physical Review* **76** (8): 1262. Bibcode:1949PhRv...76.1262G. doi:10.1103/PhysRev.76.1262.2.

[34] Carver, T. R.; Slichter, C. P. (1953). "Polarization of Nuclear Spins in Metals".*Physical Review***92**(1): 212–213.Bibcode:. doi:10.1103/PhysRev.92.212.2.

[35] Hugh Everett Theory of the Universal Wavefunction, Thesis, Princeton University, (1956, 1973), pp 1–140

[36] Everett, Hugh (1957). "Relative State Formulation of Quantum Mechanics". *Reviews of Modern Physics* **29** (3): 454–462. Bibcode:1957RvMP...29..454E. doi:10.1103/RevModPhys.29.454.

[37] Jacek W. Hennel, Jacek Klinowski (2005). Jacek Klinowski, ed. "New techniques in solid-state NMR". Topics in Current Chemistry **246**. Springer. pp. 1–14. doi:10.1007/b98646. ISBN 3-540-22168-9. |chapter= ignored (help) (*New techniques in solid-state NMR*, p. 1, at Google Books)

[38] V.E. Barnes; Connolly, P.; Crennell, D.; Culwick, B.; Delaney, W.; Fowler, W.; Hagerty, P.; Hart, E.; Horwitz, N.; Hough, P.; Jensen, J.; Kopp, J.; Lai, K.; Leitner, J.; Lloyd, J.; London, G.; Morris, T.; Oren, Y.; Palmer, R.; Prodell, A.; Radojičić, D.; Rahm, D.; Richardson, C.; Samios, N.; Sanford, J.; Shutt, R.; Smith, J.; Stonehill, D.; Strand, R.; et al. (1964). "Observation of a Hyperon with Strangeness Number Three" (PDF). *Physical Review Letters* **12** (8): 204. Bibcode:1964PhRvL..12..204B. doi:10.1103/PhysRevLett.12.204.

[39] Anatole Abragam (1961). *The Principles of Nuclear Magnetism.* Oxford: Clarendon Press. OCLC 242700.

[40] Brian David Josephson (December 12, 1973). "The Discovery of Tunnelling Supercurrents" (PDF). *Nobel Lecture.* Retrieved April 2012.

[41] Maria Goeppert Mayer (December 12, 1963). "The shell model" (PDF). *Nobel Lecture.* Retrieved April 2012.

[42] F. Englert, R. Brout; Brout (1964). "Broken Symmetry and the Mass of Gauge Vector Mesons". *Physical Review Letters* **13** (9): 321–323. Bibcode:1964PhRvL..13..321E. doi:10.1103/PhysRevLett.13.321.

[43] P.W. Higgs (1964). "Broken Symmetries and the Masses of Gauge Bosons". *Physical Review Letters* **13** (16): 508–509. Bibcode:1964PhRvL..13..508H. doi:10.1103/PhysRevLett.13.508.

[44] G.S. Guralnik, C.R. Hagen, T.W.B. Kibble; Hagen; Kibble (1964). "Global Conservation Laws and Massless Particles". *Physical Review Letters* **13** (20): 585–587. Bibcode:1964PhRvL..13..585G. doi:10.1103/PhysRevLett.13.585.

[45] G.S. Guralnik (2009). "The History of the Guralnik, Hagen and Kibble development of the Theory of Spontaneous Symmetry Breaking and Gauge Particles". *International Journal of Modern Physics A* **24** (14): 2601–2627. arXiv:0907.3466. Bibcode:2009IJMPA..24.2601G. doi:10.1142/S0217751X09045431.

[46] T.W.B. Kibble (2009)."Englert–Brout–Higgs–Guralnik–Hagen–Kibble mechanism".*Scholarpedia***4**(1): 6441.Bibcode:2009 doi:10.4249/scholarpedia.6441.

[47] M. Blume, S. Brown, Y. Millev (2008). "Letters from the past, a PRL retrospective (1964)". Physical Review Letters. Retrieved 2010-01-30.

[48] "J. J. Sakurai Prize Winners". American Physical Society. 2010. Retrieved 2010-01-30.

[49] "Discovery of the Charmed Baryon". *Brookhaven History.* Brookhaven National Laboratory.

[50] Wilczek, Frank (1999). "Quantum field theory".*Reviews of Modern Physics***71**(2): S85.arXiv:hep-th/9803075.Bibcode: doi:10.1103/RevModPhys.71.S85.

[51] Mansfield, P; Grannell, P K (1973). "NMR 'diffraction' in solids?". *Journal of Physics C: Solid State Physics* **6** (22): L422. Bibcode:1973JPhC....6L.422M. doi:10.1088/0022-3719/6/22/007.

[52] Garroway, A N; Grannell, P K; Mansfield, P (1974). "Image formation in NMR by a selective irradiative process". *Journal of Physics C: Solid State Physics* **7** (24): L457. Bibcode:1974JPhC....7L.457G. doi:10.1088/0022-3719/7/24/006.

[53] Mansfield, P.; Maudsley, A. A. (1977). "Medical imaging by NMR".*British Journal of Radiology***50**(591): 188–94.doi:10.1 1285-50-591-188. PMID 849520.

[54] Mansfield, P (1977). "Multi-planar image formation *using NMR* spin echoes". *Journal of Physics C: Solid State Physics* **10** (3): L55. Bibcode:1977JPhC...10L..55M. doi:10.1088/0022-3719/10/3/004.

[55] Ilya Prigogine (8 December 1977). "Time, Structure and Fluctuations" (PDF). *Nobel lecture.* Retrieved April 2012.

[56] Rubinson, K.A.; Rubinson, Kenneth A.; Patterson, John (1979). "Ferromagnetic resonance and spin wave excite journals in metallic glasses". *J. Phys. Chem. Solids* **40** (12): 941–950. Bibcode:1979JPCS...40..941B. doi:10.1016/0022-3697(79)90122-7.

[57] Aspect, Alain; Grangier, Philippe; Roger, Gérard (1982). "Experimental Realization of Einstein-Podolsky-Rosen-Bohm Ged anken-experiment:A New Violation of Bell's Inequalities". *Physical Review Letters*49(2):91. Bibcode:1982PhRvL..49...91A. doi:10.1103/PhysRevLett.49.91.

[58] Aspect, Alain; Dalibard, Jean; Roger, Gérard (1982). "Experimental Test of Bell's Inequalities Using Time- Varying Analyzers". *Physical Review Letters* **49** (25): 1804. Bibcode:1982PhRvL..49.1804A. doi:10.1103/PhysRevLett.49.1804.

[59] TFTR Machine Parameters. W3.pppl.gov (1996-05-10). Retrieved on 2012-05-17.

[60] JET's Main Features-EFDA JET. Jet.efda.org. Retrieved on 2012-05-17.

[61] European JET website. (PDF) . Retrieved on 2012-05-17.

[62] Japan Atomic Energy Agency. Naka Fusion Institute

[63] Fusion Plasma Research (FPR), JASEA, Naka Fusion Institute. Jt60.naka.jaea.go.jp. Retrieved on 2012-05-17.

[64] Müller, KA; Bednorz, JG (1987). "The discovery of a class of high-temperature superconductors". *Science* **237** (4819): 1133–9. Bibcode:1987Sci...237.1133M. doi:10.1126/science.237.4819.1133. PMID 17801637.

[65] Pont, M.; Walet, N.R.; Gavrila, M.; McCurdy, C.W. (1988). "Dichotomy of the Hydrogen Atom in Superintense, High-Frequency Laser Fields". *Physical Review Letters* **61** (8): 939–942. Bibcode:1988PhRvL..61..939P. doi:10.1103/PhysRevLett.61.939.PMID10039473.

[66] Pont, M.; Walet, N.; Gavrila, M. (1990). "Radiative distortion of the hydrogen atom in superintense, high-frequency fields of linear polarization". *Physical Review A* **41** (1): 477–494. Bibcode:1990PhRvA..41..477P. doi:10.1103/PhysRevA.41.477. PMID 9902891.

[67] Mihai Gavrila: *Atomic Structure and Decay in High-Frequency Fields*, in *Atoms in Intense Laser Fields*, ed. M. Gavrila, Academic Press, San Diego, 1992, pp. 435–510. ISBN 0-12-003901-X

[68] Muller, H.; Gavrila, M. (1993). "Light-Induced Excited States in H⁻". *Physical Review Letters* **71** (11): 1693–1696. Bibcode:1993 doi:10.1103/PhysRevLett.71.1693. PMID 10054474. PhRvL..71.1693M.

[69] Wells, J.C.; Simbotin, I.; Gavrila, M. (1998). "Physical Reality of Light-Induced Atomic States". *Physical Review Letters* **80** (16): 3479–3482. Bibcode:1998PhRvL..80.3479W. doi:10.1103/PhysRevLett.80.3479.

[70] Ernst, E; van Duijn, M. Gavrila; Muller, H.G. (1996). "Multiply Charged Negative Ions of Hydrogen Induced by Superintense Laser Fields". *Physical Review Letters* **77** (18): 3759–3762. Bibcode:1996PhRvL..77.3759V. doi:10.1103/PhysRevLett.77.3759. PMID 10062301.

[71] Shertzer, J.; Chandler, A.; Gavrila, M. (1994). "H₂⁺ in Superintense Laser Fields: Alignment and Spectral Restructuring". *Physical Review Letters* **73** (15): 2039–2042. Bibcode:1994PhRvL..73.2039S. doi:10.1103/PhysRevLett.73.2039. PMID 10056956.

[72] Richard R. Ernst (December 9, 1992). "Nuclear Magnetic Resonance Fourier Transform (2D-FT) Spectroscopy" (PDF). *Nobel Lecture*. Retrieved April 2012.

[73] PPPL, Princeton, USA. Pppl.gov (1999-02-12). Retrieved on 2012-05-17.

[74] "Lene Hau". Physicscentral.com. Retrieved 2013-01-30.

[75] Leonid Vainerman (2003). *Locally Compact Quantum Groups and Groupoids: Proceedings of the Meeting of Theoretical Physicists and Mathematicians, Strasbourg, February 21–23, 2002*. Walter de Gruyter. pp. 247–. ISBN 978-3-11-020005-8. Retrieved 17 May 2012.

[76] LTX EXperiment Achieves First Plasma (at PPPL). Pppl.gov. Retrieved on 2012-05-17.

[77]Aspect, A. (2007). "To be or not to be local".*Nature***446**(7138): 866–867.Bibcode:2007Natur.446..866A.doi:10.1038/44686. PMID 17443174.

[78] "Coherent Population". Defense Procurement News. 2010-06-22. Retrieved 2013-01-30.

[79] Markoff, John (29 May 2014). "Scientists Report Finding Reliable Way to Teleport Data". *New York Times*. Retrieved 29 May 2014.

[80] Pfaff, W.; et al. (29 May 2014). "Unconditional quantum teleportation between distant solid-state quantum bits". *Science (journal)*. arXiv:1404.4369. Bibcode:2014Sci...345..532P. doi:10.1126/science.1253512. Retrieved 29 May 2014.

16.6 Bibliography

- Peacock, Kent A. (2008). "The quantum revolution : a historical perspective". Westport, Conn.: Greenwood Press. ISBN 9780313334481.

- Ben-Menahem, A. (2009). "Historical encyclopedia of natural and mathematical sciences" (1st ed.). Berlin: Springer. pp. 4342–4349. ISBN 9783540688310. |chapter= ignored (help)

Electron microscope constructed by Ernst Ruska in1933.

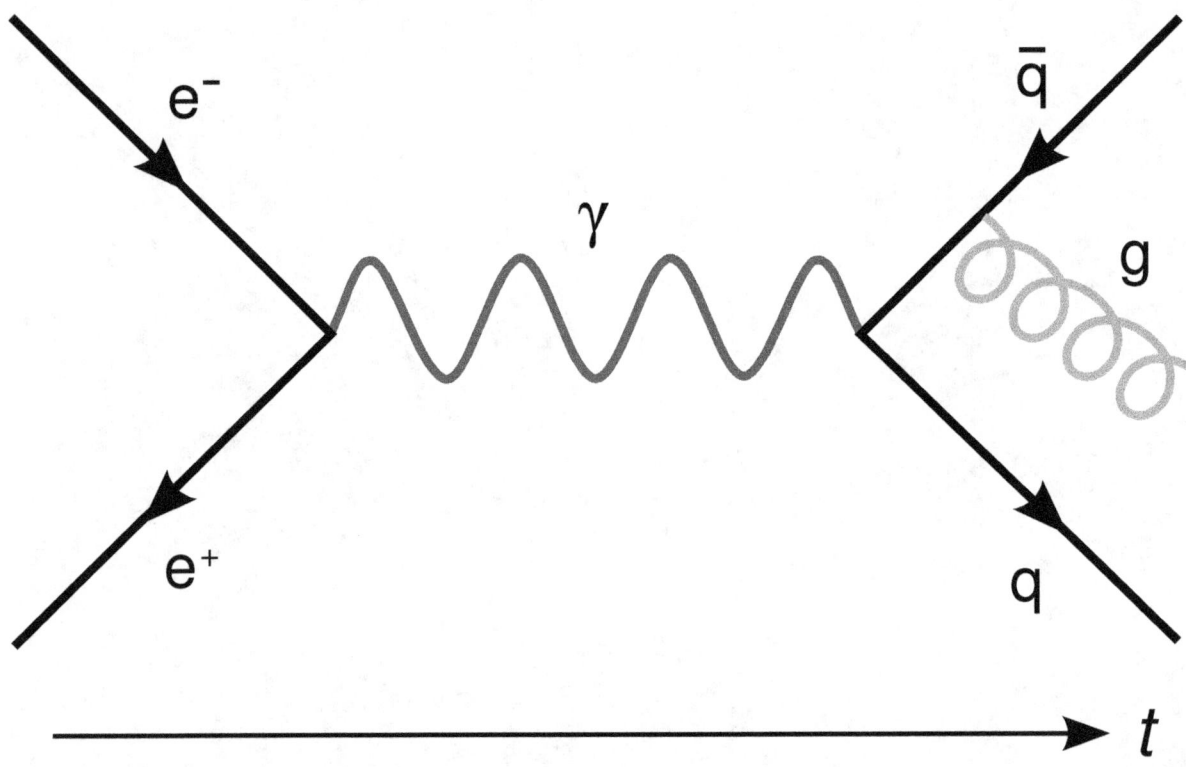

A Feynman diagram showing the radiation of a gluon when an electron and positron are annihilated.

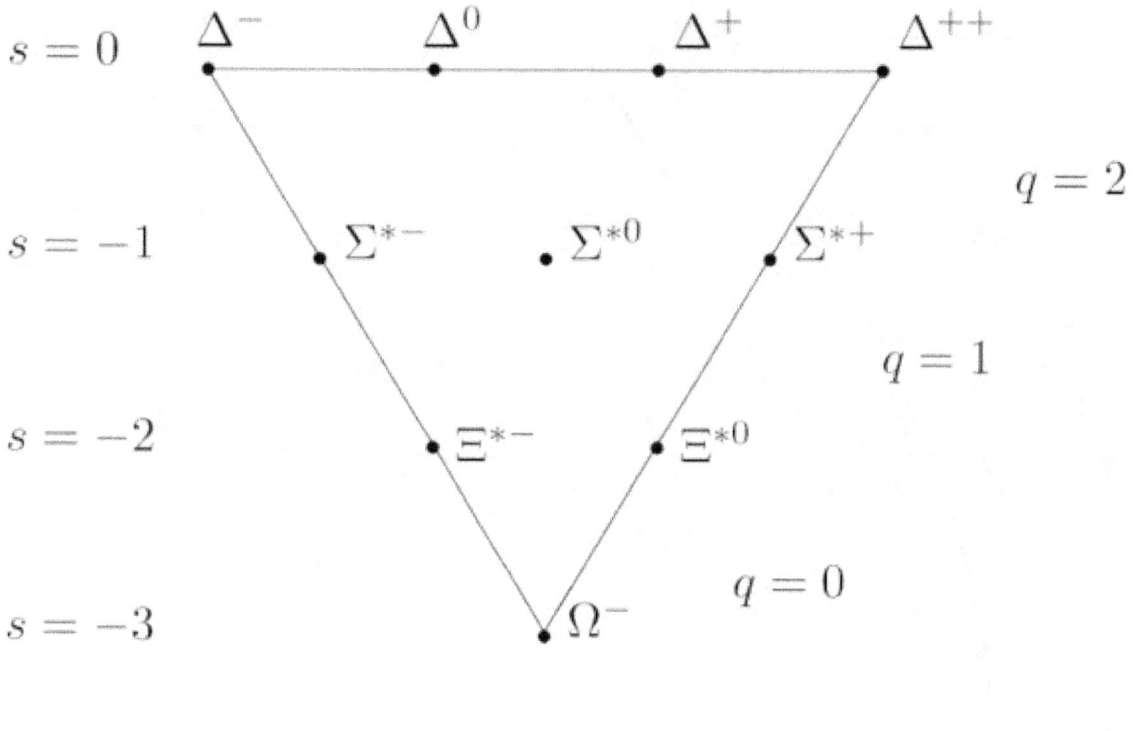

The baryon decuplet of the Eightfold Way proposed by Murray Gell-Mann in 1962. The Ω− particle at the bottom had not yet been observed at the time, but a particle closely matching these predictions was discovered[38] by a particle accelerator group at Brookhaven, proving Gell-Mann's theory.

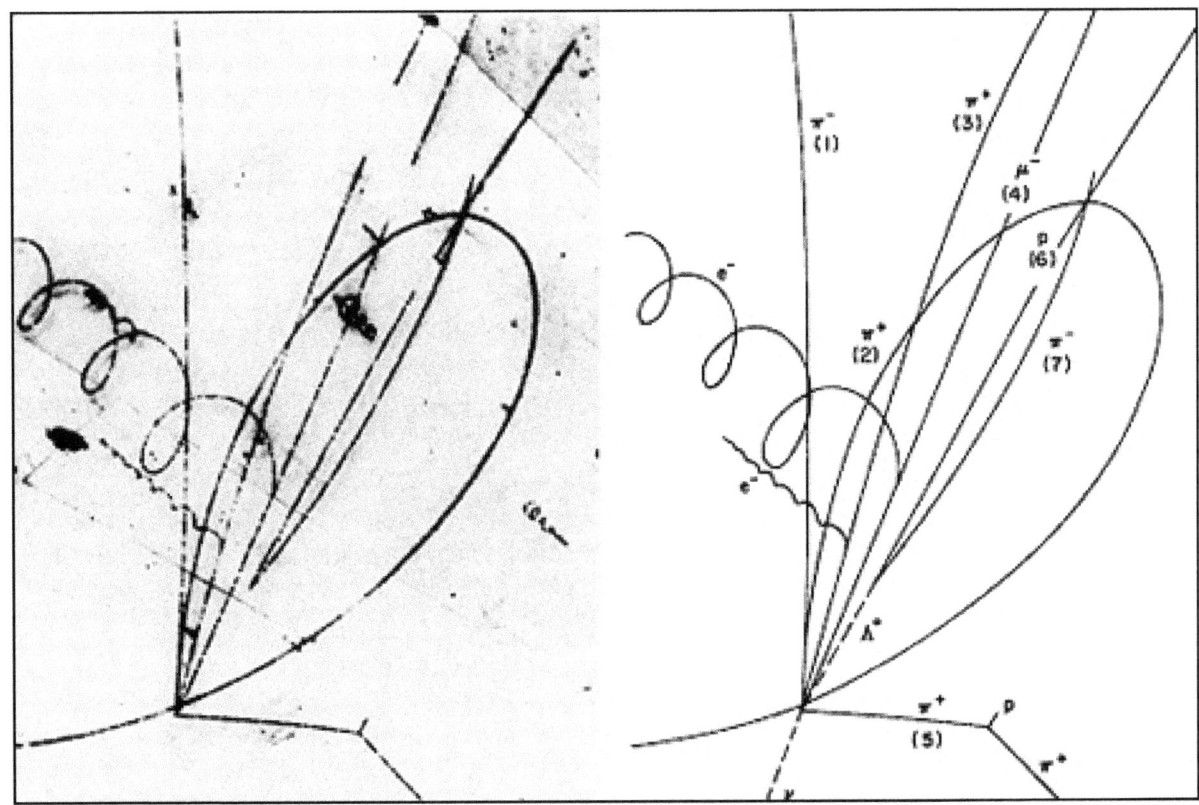

A 1974 photograph of an event in a bubble chamber at Brookhaven National Laboratory. Each track is left by a charged particle, one of which is a baryon containing the charm quark.[49]

Graphene is a planar atomic-scale honeycomb lattice made of carbon atoms which exhibits unusual and interesting quantum properties.

Chapter 17

Timeline of computational physics

Timeline of computational physics

17.1 1930s

- John Vincent Atanasoff and Clifford Berry create the first electronic non-programmable, digital computing device, the Atanasoff–Berry Computer, from 1937 to 1942.

17.2 1940s

- Nuclear bomb and ballistics simulations at Los Alamos and BRL, respectively.[1]

- Monte Carlo simulation (voted one of the top 10 algorithms of the 20th century) invented at Los Alamos by von Neumann, Ulam and Metropolis.[2][3][4]

- First hydro simulations at Los Alamos occurred.[5][6]

- Ulam and von Neumann introduce the notion of cellular automata.[7]

17.3 1950s

- *Equations of State Calculations by Fast Computing Machines* introduces the Metropolis–Hastings algorithm.[8] Also, important earlier independent work by Alder and S. Frankel.[9][10]

- Fermi, Ulam and Pasta with help from Mary Tsingou, discover the Fermi–Pasta–Ulam problem.[11]

- Molecular dynamics invented by Alder and Wainwright[12]

17.4 1960s

- Molecular dynamics was invented independently by Aneesur Rahman.[13]

- Kruskal and Zabusky follow up the Fermi–Pasta–Ulam problem with further numerical experiments, and coin the term "soliton".[14][15]

- Edward Lorenz discovers the butterfly effect on a computer, attracting interest in chaos theory.[16]

- W Kohn instigates the development of density functional theory (with LJ Sham and P Hohenberg),[17][18] for which he shares the Nobel Chemistry Prize (1998).[19] This contribution is arguably the first Nobel given for a computer programme or computational technique.

- Frenchman Verlet (re)discovers a numerical integration algorithm,[20] (first used in 1791 by Delambre, by Cowell and Crommelin in 1909, and by Carl Fredrik Störmer in 1907,[21] hence the alternative names Störmer's method or the Verlet-Störmer method) for dynamics, and the Verlet list.[20]

17.5   1970s

- Veltman's calculations at CERN lead him and t'Hooft to valuable insights into renormalizability of electroweak theory.[22] The computation has been cited as a key reason to the award of the Nobel prize to both.[23]

- Hardy, Pomeau and de Pazzis introduced the first lattice gas model, abbreviated as the HPP model after its authors.[24][25] These later evolve into lattice Boltzmann models.

- Wilson shows that continuum QCD is recovered for an infinitely large lattice with its sites infinitesimally close to one another, thereby beginning lattice QCD.[26]

17.6   1980s

- Italian physicists Car and Parrinello invent the Car–Parrinello method.[27]

- Fast multipole method invented by Rokhlin and Greengard (voted one of the top 10 algorithms of the 20th century).[28][29][30]

17.7   See also

- Timeline of scientific computing

- Computational physics

- Important publications in computational physics

17.8   References

[1] Ballistic Research Laboratory, Aberdeen Proving Grounds, Maryland.

[2] Metropolis, N. (1987). "The Beginning of the Monte Carlo method" (PDF). *Los Alamos Science*. No. 15, Page 125.. Accessed 5 may 2012.

[3] S. Ulam, R. D. Richtmyer, and J. von Neumann(1947). Statistical methods in neutron diffusion. Los Alamos Scientific Laboratory report LAMS–551.

[4] N. Metropolis and S. Ulam (1949). The Monte Carlo method. Journal of the American Statistical Association 44:335–341.

[5] Richtmyer, R. D. (1948). Proposed Numerical Method for Calculation of Shocks. Los Alamos, NM: Los Alamos Scientific Laboratory LA-671.

[6] A Method for the Numerical Calculation of Hydrodynamic Shocks. Von Neumann, J.; Richtmyer, R. D. Journal of Applied Physics, Vol. 21, pp. 232–237

[7] Von Neumann, J., Theory of Self-Reproduiing Automata, Univ. of Illinois Press, Urbana, 1966.

[8] Metropolis, N.; Rosenbluth, A.W.; Rosenbluth, M.N.; Teller, A.H.; Teller, E. (1953). "Equations of State Calculations by Fast Computing Machines".*Journal of Chemical Physics***21**(6): 1087–1092.Bibcode:1953JChPh..21.1087M.doi:10.1063/1.1699

[9] Unfortunately, Alder's thesis advisor was unimpressed, so Alder and Frankel delayed publication of their results until much later. Alder, B. J. , Frankel, S. P. , and Lewinson, B. A. , J. Chem. Phys., 23, 3 (1955).

[10] http://www.hp9825.com/html/stan_frankel.html

[11] Fermi, E. (posthumously); Pasta, J.; Ulam, S. (1955) : Studies of Nonlinear Problems (accessed 25 Sep 2012). Los Alamos Laboratory Document LA-1940. Also appeared in 'Collected Works of Enrico Fermi', E. Segre ed. , University of Chicago Press, Vol.II,978–988,1965. Recovered 21 Dec 2012

[12] Alder, B. J.; Wainwright, T. E. (1959). "Studies in Molecular Dynamics. I. General Method". *Journal of Chemical Physics* **31** (2): 459. Bibcode:1959JChPh..31..459A. doi:10.1063/1.1730376.

[13] Rahman, A (1964). "Correlations in the Motion of Atoms in Liquid Argon".*Phys Rev***136**(2A): A405–A41.Bibcode:1964PhRv doi:10.1103/PhysRev.136.A405.

[14] Zabusky, N. J.; Kruskal, M. D. (1965). "Interaction of 'solitons' in a collisionless plasma and the recurrence of initial states". Phys. Rev. Lett. 15 (6): 240–243. Bibcode 1965PhRvL..15..240Z. doi:10.1103/PhysRevLett.15.240.

[15] http://www.merriam-webster.com/dictionary/soliton ; retrieved 3 nov 2012.

[16] Lorenz, Edward N. (1963). "Deterministic Nonperiodic Flow" (PDF). *Journal of the Atmospheric Sciences 20 (2): 130–141* **20** (2): 130. Bibcode:1963JAtS...20..130L. doi:10.1175/1520-0469(1963)020<0130:DNF>2.0.CO;2.

[17] Kohn, Walter; Hohenberg, Pierre (1964)."Inhomogeneous Electron Gas".*Physical Review***136**(3B): B864–B871.Bibcode:1964 doi:10.1103/PhysRev.136.B864.

[18] Kohn, Walter; Sham, Lu Jeu (1965). "Self-Consistent Equations Including Exchange and Correlation Effects". *Physical Review* **140** (4A): A1133–A1138. Bibcode:1965PhRv..140.1133K. doi:10.1103/PhysRev.136.B864.

[19] "The Nobel Prize in Chemistry 1998". Nobelprize.org. Retrieved 2008-10-06.

[20] Verlet, Loup (1967). "Computer "Experiments" on Classical Fluids. I. Thermodynamical Properties of Lennard–Jones Molecules". *Physical Review* **159**: 98–103. Bibcode:1967PhRv..159...98V. doi:10.1103/PhysRev.159.98.

[21] Press, WH; Teukolsky, SA; Vetterling, WT; Flannery, BP (2007). "Section 17.4. Second-Order Conservative Equations". *Numerical Recipes: The Art of Scientific Computing* (3rd ed.). New York: Cambridge University Press. ISBN 978-0-521-88068-8.

[22] Frank Close. The Infinity Puzzle, pg 207. OUP, 2011.

[23] Stefan Weinzierl:- "Computer Algebra in Particle Physics." pgs 5–7. arXiv:hep-ph/0209234. All links accessed 1 January 2012. "Seminario Nazionale di Fisica Teorica", Parma, September 2002.

[24] J. Hardy, Y. Pomeau, and O. de Pazzis (1973). "Time evolution of two-dimensional model system I: invariant states and time correlation functions". *Journal of Mathematical Physics*, **14**:1746–1759.

[25] J. Hardy, O. de Pazzis, and Y. Pomeau (1976). "Molecular dynamics of a classical lattice gas: Transport properties and time correlation functions". *Physics Review A*, **13**:1949–1961.

[26] Wilson, K.(1974). "Confinement of quarks".*Physical Review D***10**(8): 2445.Bibcode:1974PhRvD..10.2445W.doi:10.1103/Ph

[27] Car, R.; Parrinello, M (1985). "Unified Approach for Molecular Dynamics and Density-Functional Theory". *Physical Review Letters* **55** (22): 2471–2474. Bibcode:1985PhRvL..55.2471C. doi:10.1103/PhysRevLett.55.2471. PMID 10032153.

[28] L. Greengard, The Rapid Evaluation of Potential Fields in Particle Systems, MIT, Cambridge, (1987).

[29] Rokhlin, Vladimir (1985). "Rapid Solution of Integral Equations of Classic Potential Theory." J. Computational Physics Vol. 60, pp. 187–207.

[30] L. Greengard and V. Rokhlin, "A fast algorithm for particle simulations," J. Comput. Phys., 73 (1987), no. 2, pp. 325–348.

17.9   External links

- The Monte Carlo Method: Classic Papers

- Monte Carlo Landmark Papers

Chapter 18

Timeline of cosmological theories

This **timeline of cosmological theories** and discoveries is a chronological record of the development of humanity's understanding of the cosmos over the last two-plus millennia. Modern cosmological ideas follow the development of the scientific discipline of physical cosmology.

18.1 Pre-1900

- ca. **16th century BCE** — Mesopotamian cosmology has a flat, circular Earth enclosed in a cosmic ocean.[1]

- ca. **12th century BCE** — The *Rigveda* has some cosmological hymns, particularly in the late book 10, notably the Nasadiya Sukta which describes the origin of the universe, originating from the monistic *Hiranyagarbha* or "Golden Egg".

- **6th century BCE** — The Babylonian world map shows the Earth surrounded by the cosmic ocean, with seven islands arranged around it so as to form a seven-pointed star. Contemporary Biblical cosmology reflects the same view of a flat, circular Earth swimming on water and overarched by the solid vault of the firmament to which are fastened the stars.

- **4th century BCE** — Aristotle proposes an Earth-centered universe in which the Earth is stationary and the cosmos (or universe) is finite in extent but infinite in time

- **4th century BCE** — De Mundo - Five elements, situated in spheres in five regions, the less being in each case surrounded by the greater — namely, earth surrounded by water, water by air, air by fire, and fire by ether — make up the whole Universe.[2]

- **3rd century BCE** — Aristarchus of Samos proposes a Sun-centered universe

- **3rd century BCE** — Archimedes in his essay The Sand Reckoner, estimates the diameter of the cosmos to be the equivalent in stadia of what we call two light years

- **2nd century BCE** — Seleucus of Seleucia elaborates on Aristarchus' heliocentric universe, using the phenomenon of tides to explain heliocentrism

- **2nd century CE** — Ptolemy proposes an Earth-centered universe, with the Sun, moon, and visible planets revolving around the Earth

- **5th-11th centuries** — Several astronomers propose a Sun-centered universe, including Aryabhata, Albumasar[3] and Al-Sijzi

- **6th century** — John Philoponus proposes a universe that is finite in time and argues against the ancient Greek notion of an infinite universe

- ca. **8th century** — Puranic Hindu cosmology, in which the Universe goes through repeated cycles of creation, destruction and rebirth, with each cycle lasting 4.32 billion years.

- **9th-12th centuries** — Al-Kindi (Alkindus), Saadia Gaon (Saadia ben Joseph) and Al-Ghazali (Algazel) support a universe that has a finite past and develop two logical arguments against the notion of an infinite past, one of which is later adopted by Immanuel Kant

- **964** — Abd al-Rahman al-Sufi (Azophi), a Persian astronomer, makes the first recorded observations of the Andromeda Galaxy and the Large Magellanic Cloud, the first galaxies other than the Milky Way to be observed from Earth, in his *Book of Fixed Stars*

- **12th century** — Fakhr al-Din al-Razi discusses Islamic cosmology, rejects Aristotle's idea of an Earth-centered universe, and, in the context of his commentary on the Qur'anic verse, "All praise belongs to God, Lord of the Worlds," proposes that the universe has more than "a thousand thousand worlds beyond this world such that each one of those worlds be bigger and more massive than this world as well as having the like of what this world has."[4] He argued that there exists an infinite outer space beyond the known world,[5] and that there could be an infinite number of universes.[6]

- **13th century** — Nasīr al-Dīn al-Tūsī provides the first empirical evidence for the Earth's rotation on its axis

- **13th century** — Nahmanides suggests the universe is expanding and that there are ten dimensions.

- **15th century** — Ali Qushji provides empirical evidence for the Earth's rotation on its axis and rejects the stationary Earth theories of Aristotle and Ptolemy

- **15th-16th centuries** — Nilakantha Somayaji and Tycho Brahe propose a universe in which the planets orbit the Sun and the Sun orbits the Earth, known as the Tychonic system

- **1543** — Nicolaus Copernicus publishes his heliocentric universe in his *De revolutionibus orbium coelestium*

- **1576** — Thomas Digges modifies the Copernican system by removing its outer edge and replacing the edge with a star-filled unbounded space

- **1584** — Giordano Bruno proposes a non-hierarchical cosmology, wherein the Copernican solar system is not the center of the universe, but rather, a relatively insignificant star system, amongst an infinite multitude of others

- **1610** — Johannes Kepler uses the dark night sky to argue for a finite universe

- **1687** — Sir Isaac Newton's laws describe large-scale motion throughout the universe

- **1720** — Edmund Halley puts forth an early form of Olbers' paradox

- **1729** - James Bradley discovers the aberration of light, due to the Earth's motion around the Sun.

- **1744** — Jean-Philippe de Cheseaux puts forth an early form of Olbers' paradox

- **1755** — Immanuel Kant asserts that the nebulae are really galaxies separate from, independent of, and outside the Milky Way Galaxy; he calls them *island universes*.

- **1785** — William Herschel proposes the theory that our Sun is at or near the center of the galaxy.

- **1791** — Erasmus Darwin pens the first description of a cyclical expanding and contracting universe in his poem *The Economy of Vegetation*

- **1826** — Heinrich Wilhelm Olbers puts forth Olbers' paradox

- **1837** - Following over 100 years of unsuccessful attempts, Friedrich Bessel, Thomas Henderson and Otto Struve measure the parallax of a few nearby stars; this is the first measurement of any distances outside the solar system.

- **1848** — Edgar Allan Poe offers first correct solution to Olbers' paradox in *Eureka: A Prose Poem*, an essay that also suggests the expansion and collapse of the universe

- **1860s** - William Huggins develops astronomical spectroscopy; he shows that the Orion nebula is mostly made of gas, while the Andromeda nebula (later called Andromeda galaxy) is probably dominated by stars.

18.2   1900–1949

- **1905** — Albert Einstein publishes the Special Theory of Relativity, positing that space and time are not separate continua

- **1912** - Henrietta Leavitt discovers the period-luminosity law for Cepheid variable stars, which becomes a crucial step in measuring distances to other galaxies.

- **1915** — Albert Einstein publishes the General Theory of Relativity, showing that an energy density warps spacetime

- **1917** — Willem de Sitter derives an isotropic static cosmology with a cosmological constant, as well as an empty expanding cosmology with a cosmological constant, termed a de Sitter universe

- **1920** — The Shapley-Curtis Debate, on the distances to spiral nebulae, takes place at the Smithsonian

- **1921** — The National Research Council (NRC) published the official transcript of the Shapley-Curtis Debate

- **1922** — Vesto Slipher summarizes his findings on the spiral nebulae's systematic redshifts

- **1922** — Alexander Friedmann finds a solution to the Einstein field equations which suggests a general expansion of space

- **1923** — Edwin Hubble measures distances to a few nearby spiral nebulae (galaxies), the Andromeda Galaxy (M31), Triangulum Galaxy (M33), and NGC 6822. The distances place them far outside our Milky Way, and implies that fainter galaxies are much more distant, and the universe is composed of many thousands of galaxies.

- **1927** — Georges Lemaître discusses the creation event of an expanding universe governed by the Einstein field equations. From its solutions to the Einstein equations, he predicts the distance-redshift relation.

- **1928** — Howard P. Robertson briefly mentions that Vesto Slipher's redshift measurements combined with brightness measurements of the same galaxies indicate a redshift-distance relation

- **1929** — Edwin Hubble demonstrates the linear redshift-distance relation and thus shows the expansion of the universe

- **1933** — Edward Milne names and formalizes the cosmological principle

- **1933** — Fritz Zwicky shows that the Coma cluster of galaxies contains large amounts of dark matter. This result agrees with modern measurements, but is generally ignored until the 1970s.

- **1934** — Georges Lemaître interprets the cosmological constant as due to a vacuum energy with an unusual perfect fluid equation of state

- **1938** — Paul Dirac suggests the large numbers hypothesis, that the gravitational constant may be small because it is decreasing slowly with time

- **1948** — Ralph Alpher, Hans Bethe ("in absentia"), and George Gamow examine element synthesis in a rapidly expanding and cooling universe, and suggest that the elements were produced by rapid neutron capture

- **1948** — Hermann Bondi, Thomas Gold, and Fred Hoyle propose steady state cosmologies based on the perfect cosmological principle

- **1948** — George Gamow predicts the existence of the cosmic microwave background radiation by considering the behavior of primordial radiation in an expanding universe

18.3   1950–1999

- **1950** — Fred Hoyle coins the term "Big Bang", saying that it was not derisive; it was just a striking image meant to highlight the difference between that and the Steady-State model.

- **1961** — Robert Dicke argues that carbon-based life can only arise when the gravitational force is small, because this is when burning stars exist; first use of the weak anthropic principle

- **1963** - Maarten Schmidt discovers the first quasar; these soon provide a probe of the universe back to substantial redshifts.

- **1965** — Hannes Alfvén proposes the now-discounted concept of ambiplasma to explain baryon asymmetry and supports the idea of an infinite universe.

- **1965** — Martin Rees and Dennis Sciama analyze quasar source count data and discover that the quasar density increases with redshift.

- **1965** — Arno Penzias and Robert Wilson, astronomers at Bell Labs discover the 2.7 K *microwave background radiation*, which earns them the 1978 Nobel Prize in Physics. Robert Dicke, James Peebles, Peter Roll and David Todd Wilkinson interpret it as a relic from the big bang.

- **1966** — Stephen Hawking and George Ellis show that any plausible general relativistic cosmology is singular

- **1966** — James Peebles shows that the hot Big Bang predicts the correct helium abundance

- **1967** — Andrei Sakharov presents the requirements for baryogenesis, a baryon-antibaryon asymmetry in the universe

- **1967** — John Bahcall, Wal Sargent, and Maarten Schmidt measure the fine-structure splitting of spectral lines in 3C191 and thereby show that the fine-structure constant does not vary significantly with time

- **1967** — Robert Wagoner, William Fowler, and Fred Hoyle show that the hot Big Bang predicts the correct deuterium and lithium abundances

- **1968** — Brandon Carter speculates that perhaps the fundamental constants of nature must lie within a restricted range to allow the emergence of life; first use of the strong anthropic principle

- **1969** — Charles Misner formally presents the Big Bang horizon problem

- **1969** — Robert Dicke formally presents the Big Bang flatness problem

- **1970** — Vera Rubin and Kent Ford measure spiral galaxy rotation curves at large radii, showing evidence for substantial amounts of dark matter.

- **1973** — Edward Tryon proposes that the universe may be a large scale quantum mechanical vacuum fluctuation where positive mass-energy is balanced by negative gravitational potential energy

- **1976** — Alex Shlyakhter uses samarium ratios from the Oklo prehistoric natural nuclear fission reactor in Gabon to show that some laws of physics have remained unchanged for over two billion years

- **1977** — Gary Steigman, David Schramm, and James Gunn examine the relation between the primordial helium abundance and number of neutrinos and claim that at most five lepton families can exist.

- **1980** — Alan Guth and Alexei Starobinsky independently propose the inflationary Big Bang universe as a possible solution to the horizon and flatness problems.

- **1981** — Viacheslav Mukhanov and G. Chibisov propose that quantum fluctuations could lead to large scale structure in an inflationary universe.

- **1982** — The first CfA galaxy redshift survey is completed.

- **1982** — Several groups including James Peebles, J. Richard Bond and George Blumenthal propose that the universe is dominated by cold dark matter.

- **1983 - 1987** — The first large computer simulations of cosmic structure formation are run by Davis, Efstathiou, Frenk and White. The results show that cold dark matter produces a reasonable match to observations, but hot dark matter does not.

- **1988** — The CfA2 Great Wall is discovered in the CfA2 redshift survey.

- **1988** — Measurements of galaxy large-scale flows provide evidence for the Great Attractor.

- **1990** — Preliminary results from NASA's COBE mission confirm the cosmic microwave background radiation has a blackbody spectrum to an astonishing one part in 10^5 precision, thus eliminating the possibility of an integrated starlight model proposed for the background by steady state enthusiasts.

- **1992** — Further COBE measurements discover the very small anisotropy of the cosmic microwave background, providing a "baby picture" of the seeds of large-scale structure when the universe was around 1/1100th of its present size and 380,000 years old.

- **1996** - The first Hubble Deep Field is released, providing a clear view of very distant galaxies when the universe was around one-third of its present age.

- **1998** — Controversial evidence for the fine structure constant varying over the lifetime of the universe is first published.

- **1998** — The Supernova Cosmology Project and High-Z Supernova Search Team discover cosmic acceleration based on distances to Type Ia supernovae, providing the first direct evidence for a non-zero cosmological constant.

- **1999** — Measurements of the cosmic microwave background radiation with finer resolution than COBE, (most notably by the BOOMERanG experiment see Mauskopf et al., 1999, Melchiorri et al., 1999, de Bernardis et al. 2000) provide evidence for oscillations (the first acoustic peak) in the anisotropy angular spectrum, as expected in the standard model of cosmological structure formation. The angular position of this peak indicates that the geometry of the universe is close to flat.

18.4 Since 2000

- **2001** — The 2dF Galaxy Redshift Survey (2dF) by an Australian/British team gave strong evidence that the matter density is near 25% of critical density. Together with the CMB results for a flat universe, this provides independent evidence for a cosmological constant or similar dark energy.

- **2002** — The Cosmic Background Imager (CBI) in Chile obtained images of the cosmic microwave background radiation with the highest angular resolution of 4 arc minutes. It also obtained the anisotropy spectrum at high-resolution not covered before up to l ~ 3000. It found a slight excess in power at high-resolution (l > 2500) not yet completely explained, the so-called "CBI-excess".

- **2003** — NASA's Wilkinson Microwave Anisotropy Probe (WMAP) obtained full-sky detailed pictures of the cosmic microwave background radiation. The image can be interpreted to indicate that the universe is 13.7 billion years old (within one percent error), and are very consistent with the Lambda-CDM model and the density fluctuations predicted by inflation.

- **2003** — The Sloan Great Wall is discovered.

- **2004** — The Degree Angular Scale Interferometer (DASI) first obtained the E-mode polarization spectrum of the cosmic microwave background radiation.

- **2005** — The Sloan Digital Sky Survey (SDSS) and 2dF redshift surveys both detected the baryon acoustic oscillation feature in the galaxy distribution, a key prediction of cold dark matter models.

- **2006** — The long-awaited three-year WMAP results are released, confirming previous analysis, correcting several points, and including polarization data.

- **2006-2011** — Improved measurements from WMAP, new supernova surveys ESSENCE and SNLS, and baryon acoustic oscillations from SDSS and WiggleZ, continue to be consistent with the standard Lambda-CDM model.

- **2014** — On March 17, 2014, astrophysicists of the BICEP2 collaboration announced the detection of inflationary gravitational waves in the B-mode power spectrum, which if confirmed, would provide clear experimental evidence for the theory of inflation.[7][8][9][10][11][12] However, on June 19, 2014, lowered confidence in confirming the cosmic inflation findings was reported.[11][13][14]

18.5   See also

18.5.1   Physical cosmology

- Chronology of the universe
  - Graphical timeline of the Big Bang
  - Graphical timeline from Big Bang to Heat Death
  - Timeline of cosmic microwave background astronomy
- List of cosmologists
- Non-standard cosmology

18.5.2   Belief systems

- Buddhist cosmology
- Jain cosmology
- Jainism and non-creationism
- Hindu cosmology
- Maya mythology

18.5.3   Others

- Cosmology@Home

18.6   References

[1] Horowitz (1998), p.xii

[2] Aristotle; Forster, E. S. (Edward Seymour), 1879-1950; Dobson, J. F. (John Frederic), 1875-1947 (1914). *De Mundo*. p. 2.

[3] "Introduction to Astronomy, Containing the Eight Divided Books of Abu Ma'shar Abalachus". *World Digital Library*. 1506. Retrieved 2013-07-16.

[4] Adi Setia (2004), "Fakhr Al-Din Al-Razi on Physics and the Nature of the Physical World: A Preliminary Survey", *Islam & Science* **2**, retrieved 2010-03-02

[5] Muammer İskenderoğlu (2002), *Fakhr al-Dīn al-Rāzī and Thomas Aquinas on the question of the eternity of the world*, Brill Publishers, p. 79, ISBN 90-04-12480-2

[6] John Cooper (1998), "al-Razi, Fakhr al-Din (1149-1209)", *Routledge Encyclopedia of Philosophy* (Routledge), retrieved 2010-03-07

[7] Staff (March 17, 2014). "BICEP2 2014 Results Release". *National Science Foundation*. Retrieved March 18, 2014.

[8] Clavin, Whitney (March 17, 2014). "NASA Technology Views Birth of the Universe". *NASA*. Retrieved March 17, 2014.

[9] Overbye, Dennis (March 17, 2014). "Space Ripples Reveal Big Bang's Smoking Gun". *The New York Times*. Retrieved March 17, 2014.

[10] Overbye, Dennis (March 24, 2014). "Ripples From the Big Bang". *New York Times*. Retrieved March 24, 2014.

[11] Ade, P.A.R.; BICEP2 Collaboration (June 19, 2014). "Detection of B-Mode Polarization at Degree Angular Scales by BICEP2" (PDF). *Physical Review Letters* **112**: 241101. arXiv:1403.3985. Bibcode:2014PhRvL.112x1101A. doi:10.1103/PhysRevLett.112.241101.PMID24996078.Retrieved June20,2014.

[12] http://www.math.columbia.edu/~{}woit/wordpress/?p=6865

[13] Overbye, Dennis (June 19, 2014). "Astronomers Hedge on Big Bang Detection Claim". *New York Times*. Retrieved June 20, 2014.

[14] Amos, Jonathan (June 19, 2014). "Cosmic inflation: Confidence lowered for Big Bang signal". *BBC News*. Retrieved June 20, 2014.

- Horowitz, Wayne (1998). *Mesopotamian cosmic geography*. Eisenbrauns.

- Bunch, Bryan, and Alexander Hellemans, "*The History of Science and Technology: A Browser's Guide to the Great Discoveries, Inventions, and the People Who Made Them from the Dawn of Time to Today*". ISBN 0-618-22123-9

- P. Mauskopf et al.,astro-ph/9911444, Astrophys.J. 536 (2000) L59-L62.

- A. Melchiorri et al.,astro-ph/9911445, Astrophys.J. 536 (2000) L63-L66.

- P. de Bernardis et al., astro-ph/0004404, Nature 404 (2000) 955-959.

- A. Readhead et al., Polarization observations with the Cosmic Background Imager, Science 306 (2004), 836-844.

Chapter 19

Timeline of black hole physics

Timeline of black hole physics

- 1640 — Ismaël Bullialdus suggests an inverse-square gravitational force law

- 1676 — Ole Rømer proves that light has a finite speed

- 1684 — Isaac Newton writes down his inverse-square Law of universal gravitation

- 1758 — Rudjer Josip Boscovich develops his Theory of forces, where gravity can be repulsive on small distances. So according to him strange classical bodies, such as white holes, can exist, which won't allow other bodies to reach their surfaces

- 1784 — John Michell discusses classical bodies which have escape velocities greater than the speed of light

- 1795 — Pierre Laplace discusses classical bodies which have escape velocities greater than the speed of light

- 1798 — Henry Cavendish measures the gravitational constant G

- 1876 — William Kingdon Clifford suggests that the motion of matter may be due to changes in the geometry of space

- 1909 — Albert Einstein, together with Marcel Grossmann, starts to develop a theory which would bind metric tensor g_{ik}, which defines a space geometry, with a source of gravity, that is with mass

- 1910 — Hans Reissner and Gunnar Nordström defines Reissner-Nordström singularity, Hermann Weyl solves special case for a point-body source

- 1916 — Karl Schwarzschild solves the Einstein vacuum field equations for uncharged spherically-symmetric non-rotating systems

- 1917 — Paul Ehrenfest gives conditional principle a three-dimensional space

- 1918 — Hans Reissner and Gunnar Nordström solve the Einstein–Maxwell field equations for charged spherically-symmetric non-rotating systems

- 1918 — Friedrich Kottler gets Schwarzschild solution without Einstein vacuum field equations

- 1923 — George David Birkhoff proves that the Schwarzschild spacetime geometry is the unique spherically symmetric solution of the Einstein vacuum field equations

- 1931 — Subrahmanyan Chandrasekhar calculates, using special relativity, that a non-rotating body of electron-degenerate matter above a certain limiting mass (at 1.4 solar masses) has no stable solutions.

- 1939 — Robert Oppenheimer and Hartland Snyder calculate the gravitational collapse of a pressure-free homogeneous fluid sphere

- 1958 — David Finkelstein theorises that the Schwarzschild radius of a black holes is a causality barrier: an event horizon

- 1963 — Roy Kerr solves the Einstein vacuum field equations for uncharged symmetric rotating systems, deriving the Kerr metric

- 1964 — Roger Penrose proves that an imploding star will necessarily produce a singularity once it has formed an event horizon

- 1964 — The first recorded use of the term 'Black Hole' by a journalist Ann Ewing

- 1965 — Ezra T. Newman, E. Couch, K. Chinnapared, A. Exton, A. Prakash, and Robert Torrence solve the Einstein-Maxwell field equations for charged rotating systems

- 1967 — Werner Israel presented the proof of the no-hair theorem at King's College London

- 1967 — John Wheeler helps to popularize the term "black hole"

- 1968 — Brandon Carter uses Hamilton–Jacobi theory to derive first-order equations of motion for a charged particle moving in the external fields of a Kerr-Newman black hole

- 1969 — Roger Penrose discusses the Penrose process for the extraction of the spin energy from a Kerr black hole

- 1969 — Roger Penrose proposes the cosmic censorship hypothesis

- 1971 — Identification of Cygnus X-1/HDE 226868 as a binary black hole candidate system

- 1972 — Stephen Hawking proves that the area of a classical black hole's event horizon cannot decrease

- 1972 — James Bardeen, Brandon Carter, and Stephen Hawking propose four laws of black hole mechanics in analogy with the laws of thermodynamics

- 1972 — Jacob Bekenstein suggests that black holes have an entropy proportional to their surface area due to information loss effects

- 1974 — Stephen Hawking applies quantum field theory to black hole spacetimes and shows that black holes will radiate particles with a black-body spectrum which can cause black hole evaporation

- 1989 — Identification of GS2023+338/V404 Cygni as a binary black hole candidate system

- 2002 — Astronomers at the Max Planck Institute for Extraterrestrial Physics present evidence for the hypothesis that Sagittarius A* is a supermassive black hole at the center of the Milky Way galaxy

- 2002 — NASA's Chandra X-ray Observatory identifies double galactic black holes system in merging galaxies NGC 6240

- 2004 — Further observations by a team from UCLA present even stronger evidence supporting Sagittarius A* as a black hole.

- 2012 — First visual proof of existence of black-holes. Suvi Gezari's team in Johns Hopkins University, using the Hawaiian telescope Pan-STARRS 1, publish images of a supermassive black hole 2.7 million light-years away swallowing a red giant.[1]

19.1 References

[1] Scientific American - Big Gulp: Flaring Galaxy Marks the Messy Demise of a Star in a Supermassive Black Hole

19.2   See also

- Timeline of gravitational physics and relativity

Chapter 20

Timeline of heat engine technology

This **Timeline of heat engine technology** describes how heat engines have been known since antiquity but have been made into increasingly useful devices since the 17th century as a better understanding of the processes involved was gained. They continue to be developed today.

In engineering and thermodynamics, a heat engine performs the conversion of heat energy to mechanical work by exploiting the temperature gradient between a hot "source" and a cold "sink". Heat is transferred to the sink from the source, and in this process some of the heat is converted into work.

A heat pump is a heat engine run in reverse. Work is used to create a heat differential. The timeline includes devices classed as both engines and pumps, as well as identifying significant leaps in human understanding.

20.1   Pre Seventienth century

- Prehistory - The fire piston used by tribes in southeast Asia and the Pacific islands to kindle fire.

- c. 450 BC - Archytas of Tarentum used a jet of steam to propel a toy wooden bird suspended on wire.[1]

- c. 200 BC - Hero of Alexandria's Engine, also known as Aeolipile. Demonstrates rotary motion produced by the reaction from jets of steam.

- c. 10th century - China develops the earliest fire lances which were spear-like weapons combining a bamboo tube containing gunpowder and shrapnel like projectiles tied to a spear.

- c 12th century - China, the earliest depiction of a gun showing a metal body and a tight-fitting projectile which maximises the conversion of the hot gases to forward motion.

- 1120 - Gerbert, a professor in the schools at Rheims designed and built an organ blown by air escaping from a vessel in which it was compressed by heated water.

- 1232 - First recorded use of a rocket. In a battle between the Chinese and the Mongols. (see Timeline of rocket and missile technology for a view of rocket development through time.)

- c. 1500 - Leonardo da Vinci builds the Architonnerre, a steam-powered cannon.

- 1551 - Taqi al-Din demonstrates a steam turbine, used to rotate a spit.[2]

20.2   Seventienth century

- 1629 - Giovanni Branca demonstrates a steam turbine.

- 1662 - Robert Boyle publishes Boyle's Law which defines the relationship between volume and pressure in a gas.

- 1665 - Edward Somerset, the Second Marquess of Worcester builds a working steam fountain.

- 1680 - Christiaan Huygens publishes a design for a piston engine powered by gunpowder but it is never built.

- 1690 - Denis Papin - produces design for the first piston steam engine.

- 1698 - Thomas Savery builds a pistonless steam-powered water pump for pumping water out of mines.

20.3   Eighteenth century

- 1707 - Denis Papin - produces design for his second piston steam engine in conjunction with Gottfried Leibniz.

- 1712 - Thomas Newcomen builds a piston-and-cylinder steam-powered water pump for pumping water out of mines

- 1748 - William Cullen demonstrates the first artificial refrigeration at the University of Glasgow in Scotland.

- 1759 - John Harrison uses a bimetallic strip in his third marine chronometer (H3) to compensate for temperature-induced changes in the balance spring.

- 1769 - James Watt patents his first improved steam engine, see Watt steam engine

- 1787 - Jacques Charles formulates Charles's law which describes the relationship between as gas's volume and temperature. He does not publish this however and it is not recognised until Joseph Louis Gay-Lussac develops and references it in 1802.

- 1791 - John Barber patents the idea of a gas turbine.

- 1799 - Richard Trevithick builds the first high pressure steam engine.

20.4   Nineteenth century

- 1802 - Joseph Louis Gay-Lussac develops Gay-Lussac's law which describes the relationship between a gas's pressure and temperature.

- 1807 - Nicéphore Niépce installed his 'moss, coal-dust and resin' fuelled Pyréolophore internal combustion engine in a boat and powered up the river Saone in France.

- 1807 - Franco/Swiss engineer François Isaac de Rivaz built the De Rivaz engine, powered by the internal combustion of hydrogen and oxygen mixture and used it to power a wheeled vehicle.[3]

- 1816 - Robert Stirling invented Stirling engine, a type of hot air engine.

- 1824 - Nicolas Léonard Sadi Carnot developed the Carnot cycle and the associated hypothetical Carnot heat engine that is the basic theoretical model for all heat engines. This gives the first early insight into the second law of thermodynamics.

- 1834 - Jacob Perkins, obtained the first patent for a vapor-compression refrigeration system.

- 1850s - Rudolf Clausius sets out the concept of the thermodynamic system and positioned entropy as being that in any irreversible process a small amount of heat energy δQ is incrementally dissipated across the system boundary

- 1859 - Etienne Lenoir developed the first commercially successful internal combustion engine, a single-cylinder, two-stroke engine with electric ignition of illumination gas (not gasoline).

- 1861 - Alphonse Beau de Rochas of France originates the concept of the four-stroke internal-combustion engine by emphasizing the previously unappreciated importance of compressing the fuel–air mixture before ignition.

- 1861 - Nikolaus Otto patents a two-stroke internal combustion engine building on Lenoir's.

- 1872 - Pulsometer steam pump, a pistonless pump, patented by Charles Henry Hall. It was inspired by the Savery steam pump.

- 1873 - The British chemist Sir William Crookes invents the light mill a device which turns the radiant heat of light directly into rotary motion.

- 1877 - Theorist Ludwig Boltzmann visualized a probabilistic way to measure the entropy of an ensemble of ideal gas particles, in which he defined entropy to be proportional to the logarithm of the number of microstates such a gas could occupy.

- 1877 - Nikolaus Otto patents a practical four-stroke internal combustion engine (U.S. Patent 194,047)

- 1883 - Samuel Griffin of Bath UK patents a six-stroke internal combustion engine.[4]

- 1884 - Charles A. Parsons builds the first modern Steam turbine.

- 1886 - Herbert Akroyd Stuart builds the prototype Hot bulb engine, an oil fueled Homogeneous Charge Compression Ignition engine similar to the later diesel but with a lower compression ratio and running on a fuel air mixture.

- 1892 - Rudolf Diesel patents the Diesel engine (U.S. Patent 608,845) where a high compression ratio generates hot gas which then ignites an injected fuel.

20.5 Twentieth century

- 1909, the Dutch physicist Heike Kamerlingh Onnes develops the concept of enthalpy for the measure of the "useful" work that can be obtained from a closed thermodynamic system at a constant pressure.

- 1913 - Nikola Tesla patents the Tesla turbine based on the Boundary layer effect.

- 1926 - Robert Goddard of the USA launches the first liquid fuel rocket.

- 1929 - Felix Wankel patents the Wankel rotary engine (U.S. Patent 2,988,008)

- 1933 - French physicist Georges J. Ranque invents the Vortex tube, a fluid flow device without moving parts, that can separate a compressed gas into hot and cold streams.

- 1937 - Hans von Ohain builds a gas turbine

- 1940 - Hungarian Bela Karlovitz working for the Westinghouse company in the USA files the first patent for a magnetohydrodynamic generator, which can generate electricity directly from a hot moving gas

- 1942 - R.S. Gaugler of General Motors patents the idea of the Heat pipe, a heat transfer mechanism that combines the principles of both thermal conductivity and phase transition to efficiently manage the transfer of heat between two solid interfaces.

- 1950s - The Philips company develop the Stirling-cycle Stirling Cryocooler which converts mechanical energy to a temperature difference.

- 1962 - William J. Buehler and Frederick Wang discover the Nickel titanium alloy known as Nitinol which has a shape memory dependent on its temperature.

- 1992 - The first practical magnetohydrodynamic generators are built in Serbia and the USA.

20.6   Twenty-first century

- 2011 - Michigan State University builds the first wave disk engine. An internal combustion engine which does away with pistons, crankshafts and valves, and replaces them with a disc-shaped shock wave generator.[5]

20.7   See also

- Timeline of rocket and missile technology - Rockets can be considered to be heat engines. The heat of their exhaust gases is converted into mechanical energy.

- History of thermodynamics

- History of the internal combustion engine

- Timeline of motor and engine technology

- Timeline of steam power

- Timeline of temperature and pressure measurement technology

20.8   References

- *The Growth Of The Steam-Engine* Robert H. Thurston, A. M., C. E., New York: D. Appleton and Company, 1878.

- Thermal Engineering in Power Systems By Ryoichi Amano, Bengt Sundén, Page 40, chapter 'Brief History of energy conversion'. Volume 22 of Developments in Heat Transfer Series, International series on developments in heat transfer, v. 22, WIT Press, 2008 ISBN 1-84564-062-4, ISBN 978-1-84564-062-0

20.9   Notes

[1] Hellemans, Alexander; et al. (1991). ""The Timetables of Science: A Chronology of the Most Important People and Events in the History of Science"". New York: Touchstone/Simon & Schuster, Inc., 1991.

[2] Hassan, Ahmad Y. "Taqi al-Din and the First Steam Turbine". *History of Science and Technology in Islam*. Retrieved 2008-03-29.

[3] "The History of the Automobile - Gas Engines". About.com. 2009-09-11. Retrieved 2009-10-19.

[4] The Griffin Engineering Company, of Bath, Somerset University Of Bath, 15 December 2004. Accessed May 2011

[5] Michigan State University: Wave Disk Engine U.S. Department of Energy , Advanced Research Projects Agency, March 2011

Chapter 21

Timeline of particle physics technology

Timeline of particle physics technology

- 1896 - Charles Wilson discovers that energetic particles produce droplet tracks in supersaturated gases

- 1897-1901 - Discovery of the Townsend discharge by John Sealy Townsend

- 1908 - Hans Geiger and Ernest Rutherford use the Townsend discharge principle to detect alpha particles.

- 1911 - Charles Wilson finishes a sophisticated cloud chamber

- 1928 - Hans Geiger and Walther Muller invent the Geiger Muller tube, which is based upon the gas ionisation principle used by Geiger in 1908, but is a practical device that can also detect beta and gamma radiation. This is implicitly also the invention of the Geiger Muller counter.

- 1934 - Ernest Lawrence and Stan Livingston invent the cyclotron

- 1945 - Edwin McMillan devises a synchrotron

- 1952 - Donald Glaser develops the bubble chamber

- 1968 - Georges Charpak and Roger Bouclier build the first multiwire proportional mode particle detection chamber

Chapter 22

Timeline of physical chemistry

The **timeline of physical chemistry** lists the sequence of physical chemistry theories and discoveries in chronological order.

Date	Person	Contribution
1088	Shen Kuo	First person to write of the magnetic needle compass and that it improved the accuracy of navigation by helping to employ the astronomical concept of True North at all times of the day, thus making the first, recorded, scientific observation of the magnetic field (as opposed to a theory grounded in superstition or mysticism).
1187	Alexander Neckham	First in Europe to describe the magnetic compass and its use in navigation.
1269	Pierre de Maricourt	Published the first extant treatise on the properties of magnetism and compass needles.
1550	Gerolamo Cardano	Wrote about electricity in *De Subtilitate* distinguishing, perhaps for the first time, between electrical and magnetic forces.
1600	William Gilbert	In *De Magnete*, expanded on Cardano's work (1550) and coined the New Latin word *electricus* from ἤλεκτρον (elektron), the Greek word for "amber" (from which the ancients knew an electric spark could be created by rubbing it with silk). Gilbert undertook a number of careful electrical experiments, in the course of which he discovered that many substances other than amber, such as sulphur, wax, glass, etc., were capable of manifesting electrostatic properties. Gilbert also discovered that a heated body lost its electricity and that moisture prevented the electrification of all bodies, due to the now well-known fact that moisture impairs the electrical insulation of such bodies. He also noticed that electrified substances attracted all other substances indiscriminately, whereas a magnet only attracted iron. The many discoveries of this nature earned for Gilbert the title of founder of the electrical sciences.
1646	Sir Thomas Browne	The first usage of the word *electricity* is ascribed to his work Pseudodoxia Epidemica.
1660	Otto von Guericke	Invented an early electrostatic generator. By the end of the 17th Century, researchers had developed practical means of generating electricity by friction by the use of an electrostatic generator, but the development of electrostatic machines did not begin in earnest until the 18th century, when they became fundamental instruments in the studies of the new science of electricity.
1667	Johann Joachim Becher	Stated the now-defunct scientific theory that postulated the existence of a fire-like element called "phlogiston" that was contained within combustible bodies and released during combustion. The theory was an attempt to explain processes such as combustion and the rusting of metals, which are now understood as oxidation, and which was ultimately disproved by Antoine Lavoisier in 1789.
1675	Robert Boyle	Discovered that electric attraction and repulsion can act across a vacuum and does not depend upon the air as a medium. He also added resin to the then-known list of "electrics."
1678	Christiaan Huygens	Stated his theory to the French Academy of Sciences that light is a wave-like phenomenon.
1687	Sir Isaac Newton	Published *Philosophiæ Naturalis Principia Mathematica*, by itself considered to be among the most influential books in the history of science, laying the groundwork for most of classical mechanics. In this work, Newton described universal gravitation and the three laws of motion, which dominated the scientific view of the physical universe for the next three centuries. Newton showed that the motions of objects on Earth and of celestial bodies are governed by the same set of natural laws by demonstrating the consistency between Kepler's laws of planetary motion and his theory of gravitation, thus removing the last doubts about heliocentrism and advancing the scientific revolution. In mechanics, Newton enunciated the principles of conservation of both momentum and angular momentum. (*Eventually, it was determined that Newton's laws of classical mechanics were a special case of the more general theory of quantum mechanics for macroscopic objects (in the same way that Newton's laws of motion are a special case of Einstein's Theory of Relativity)*).
1704	Sir Isaac Newton	In his work *Opticks*, Newton contended that light was made up of numerous small particles. This hypothesis could explain such features as light's ability to travel in straight lines and reflect off surfaces. However, this proposed theory was known to have its problems: although it explained reflection well, its explanation of refraction and diffraction was less satisfactory. In order to explain refraction, Newton postulated an "Aethereal Medium" transmitting vibrations faster than light, by which light, when overtaken, is put into "Fits of easy Reflexion and easy Transmission", which he supposed caused the phenomena of refraction and diffraction.
1708	Brook Taylor	Obtained a remarkable solution of the problem of the "centre of oscillation" fundamental to the development of wave mechanics which, however, remained unpublished until May 1714.
1715	Brook Taylor	In *Methodus Incrementorum Directa et Inversa* (1715), he added a new branch to the higher mathematics, now designated the "calculus of finite differences." Among other ingenious applications, he used it to determine the form of movement of a vibrating string, first successfully reduced by him to mechanical principles. The same work contained the celebrated formula known as Taylor's theorem, the importance of which remained unrecognized until 1772, when J. L. Lagrange realized its powers and termed it "le principal fondement du calcul différentiel" ("the main foundation of differential calculus"). Taylor's work thereby provided the cornerstone of the calculus of wave mechanics.
1722	René Antoine Ferchault de Réaumur	Demonstrated that iron was transformed into steel through the absorption of some substance, now known to be carbon.

1729	Stephen Gray	Conducted a series of experiments that demonstrated the difference between conductors and non-conductors (insulators). From these experiments he classified substances into two categories: "electrics", like glass, resin and silk, and "non-electrics", like metal and water. Although Gray was the first to discover and deduce the property of electrical conduction, he incorrectly stated that "electrics" conducted charges while "non-electrics" held the charge.
1732	C. F. du Fay	Conducted several experiments and concluded that all objects, except metals, animals, and liquids, could be electrified by rubbing them and that metals, animals and liquids could be electrified by means of an "electric machine" (the name used at the time for electrostatic generators), thus discrediting Gray's "electrics" and "non-electrics" classification of substances (1729).
1737	C. F. du Fay and Francis Hauksbee the younger	Independently discovered what they believed to be two kinds of frictional electricity: one generated from rubbing glass, the other from rubbing resin. From this, Du Fay theorized that electricity consists of two "electrical fluids": "vitreous" and "resinous", that are separated by friction, and that neutralize each other when combined. This two-fluid theory would later give rise to the concept of positive and negative electrical charges devised by Benjamin Franklin.
1740	Jean le Rond d'Alembert	In *Mémoire sur la réfraction des corps solides*, explains the process of refraction.
1740s	Leonhard Euler	Disagreed with Newton's corpuscular theory of light in the *Opticks*, which was then the prevailing theory. His 1740s papers on optics helped ensure that the wave theory of light proposed by Christiaan Huygens would become the dominant mode of thought, at least until the development of the quantum theory of light.
1745	Pieter van Musschenbroek	At Leiden University, he invented the Leyden jar, a type of capacitor (also known as a "condensor") for electrical energy in large quantities.
1747	William Watson	While experimenting with a Leyden jar (1745), he discovered the concept of an electrical potential (voltage) when he observed that a discharge of static electricity caused the electric current earlier observed by Stephen Gray to occur.
1752	Benjamin Franklin	Identified lightning with electricity when he discovered that lightning conducted through a metal key could be used to charge a Leyden jar, thus proving that lightning was an electric discharge and current (1747). He is also attributed with the convention of using "negative" and "positive" to denote an electrical charge or potential.
1766	Henry Cavendish	The first to recognize hydrogen gas as a discrete substance, by identifying the gas from a metal-acid reaction as "flammable air" and further finding in 1781 that the gas produces water when burned.
1771	Luigi Galvani	Invented the voltaic cell. Galvani made this discovery when he noted that two different metals (copper and zinc for example) were connected together and then both touched to different parts of a nerve of a frog leg at the same time, a spark was generated which made the leg contract. Although he incorrectly assumed that the electric current was proceeding from the frog as some kind of "animal electricity", his invention of the voltaic cell was fundamental to the development of the electric battery.
1772	Antoine Lavoisier	Showed that diamonds are a form of carbon, when he burned samples of carbon and diamond then showed that neither produced any water and that both released the same amount of carbon dioxide per gram.
1772	Carl Wilhelm Scheele	Showed that graphite, which had been thought of as a form of lead, was instead a type of carbon.
1772	Daniel Rutherford	Discovered and studied nitrogen, calling it noxious air or fixed air because this gas constituted a fraction of air that did not support combustion. Nitrogen was also studied at about the same time by Carl Wilhelm Scheele, Henry Cavendish, and Joseph Priestley, who referred to it as burnt air or phlogisticated air. Nitrogen gas was inert enough that Antoine Lavoisier referred to it as "mephitic air" or azote, from the Greek word ἄζωτος (azotos) meaning "lifeless". Animals died in it, and it was the principal component of air in which animals had suffocated and flames had burned to extinction.
1772	Carl Wilhelm Scheele	Produced oxygen gas by heating mercuric oxide and various nitrates by about 1772. Scheele called the gas 'fire air' because it was the only known supporter of combustion, and wrote an account of this discovery in a manuscript he titled Treatise on Air and Fire, which he sent to his publisher in 1775. However, that document was not published until 1777.
1778	Carl Scheele and Antoine Lavoisier	Discovered that air is composed mostly of nitrogen and oxygen.
1781	Joseph Priestley	The first to utilize the electric spark to produce an explosion of hydrogen and oxygen, mixed in the proper proportions, to produce pure water.
1784	Henry Cavendish	Discovered the inductive capacity of dielectrics (insulators) and, as early as 1778, measured the specific inductive capacity for beeswax and other substances by comparison with an air condenser.
1784	Charles-Augustin de Coulomb	Devised the torsion balance, by means of which he discovered what is known as Coulomb's law: the force exerted between two small electrified bodies varies inversely as the square of the distance; not as Franz Aepinus in his theory of electricity had assumed, merely inversely as the distance.

1788	Joseph-Louis Lagrange	Stated a re-formulation of classical mechanics that combines conservation of momentum with conservation of energy, now called Lagrangian mechanics, and which would be critical to the later development of a quantum mechanical theory of matter and energy.
1789	Antoine Lavoisier	In his text *Traité Élémentaire de Chimie* (often considered to be the first modern chemistry text), stated the first version of the law of conservation of mass, recognized and named oxygen (1778) and hydrogen (1783), abolished the phlogiston theory, helped construct the metric system, wrote the first extensive list of elements, and helped to reform chemical nomenclature.
1798	Louis Nicolas Vauquelin	In 1797 received samples of crocoite ore from which he produced chromium oxide (CrO3) by mixing crocoite with hydrochloric acid. In 1798, Vauquelin discovered that he could isolate metallic chromium by heating the oxide in a charcoal oven. He was also able to detect traces of chromium in precious gemstones, such as ruby or emerald.
1798	Louis Nicolas Vauquelin	Discovered beryllium in emerald (beryl) when he dissolved the beryl in sodium hydroxide, separating the aluminium hydroxide and beryllium compound from the silicate crystals, and then dissolving the aluminium hydroxide in another alkali solution to separate it from the beryllium.
1800	William Nicholson and Johann Ritter	Used electricity to decompose water into hydrogen and oxygen, thereby discovering the process of electrolysis, which led to the discovery of many other elements.
1800	Alessandro Volta	Invented the voltaic pile, or "battery", specifically to disprove Galvani's animal electricity theory.
1801	Johann Wilhelm Ritter	Discovered ultraviolet light.
1803	Thomas Young	Double-slit experiment supports the wave theory of light and demonstrates the effect of interference.
1806	Alessandro Volta	Employing a voltaic pile of approximately 250 cells, or couples, decomposed potash and soda, showing that these substances were respectively the oxides of potassium and sodium, which metals previously had been unknown. These experiments were the beginning of electrochemistry.
1807	John Dalton	Published his *Atomic Theory of Matter*.
1807	Sir Humphry Davy	First isolates sodium from caustic soda and potassium from caustic potash by the process of electrolysis.
1808	Sir Humphry Davy, Joseph Louis Gay-Lussac, and Louis Jacques Thénard	Boron isolated through the reaction of boric acid and potassium.
1809	Sir Humphry Davy	First publicly demonstrated the electric arc light.
1811	Amedeo Avogadro	Proposed that the volume of a gas (at a given pressure and temperature) is proportional to the number of atoms or molecules, regardless of the nature of the gas—a key step in the development of the Atomic Theory of Matter.
1817	Johan August Arfwedson and Jöns Jakob Berzelius	Arfwedson, then working in the laboratory of Berzelius, detected the presence of a new element while analyzing petalite ore. This element formed compounds similar to those of sodium and potassium, though its carbonate and hydroxide were less soluble in water and more alkaline. Berzelius gave the alkaline material the name "lithos", from the Greek word λίθος (transliterated as lithos, meaning "stone"), to reflect its discovery in a solid mineral, as opposed to sodium and potassium, which had been discovered in plant tissues.
1819	Hans Christian Oersted	Discovered the deflecting effect of an electric current traversing a wire upon a suspended magnetic needle, thus deducing that magnetism and electricity were somehow related to each other.
1821	Augustin-Jean Fresnel	Demonstrated via mathematical methods that polarization could be explained only if light was *entirely* transverse, with no longitudinal vibration whatsoever. This finding was later very important to Maxwell's equations and to Einstein's Theory of Special Relativity. His use of two plane mirrors of metal, forming with each other an angle of nearly 180°, allowed him to avoid the diffraction effects caused (by the apertures) in the experiment of F. M. Grimaldi on interference. This allowed him to conclusively account for the phenomenon of interference in accordance with the wave theory. With François Arago he studied the laws of the interference of polarized rays. He obtained circularly polarized light by means of a rhombus of glass, known as a Fresnel rhomb, having obtuse angles of 126° and acute angles of 54°.
1821	André-Marie Ampère	Announced his celebrated theory of electrodynamics, relating the force one current exerts upon another by way of its electro-magnetic effects.
1821	Thomas Johann Seebeck	Discovered the thermoelectric effect.

1831	Macedonio Melloni	Used a thermopile to detect infrared radiation.
1831	Michael Faraday	Discovered electromagnetic induction, making possible the invention of the electric motor and generator.
1833	William Rowan Hamilton	Stated a reformulation of classical mechanics that arose from Lagrangian mechanics, a previous reformulation of classical mechanics introduced by Joseph-Louis Lagrange in 1788, but which can be formulated *without* recourse to Lagrangian mechanics using symplectic spaces (see *Mathematical Formalism*). As with Lagrangian mechanics, Hamilton's equations provide a new and equivalent way of looking at classical mechanics. Generally, these equations do not provide a more convenient way of solving a particular problem. Rather, they provide deeper insights into both the general structure of classical mechanics and its connection to quantum mechanics as understood through Hamiltonian mechanics, as well as its connection to other areas of science.
1833	Michael Faraday	Announced his important law of electrochemical equivalents, viz.: "The same quantity of electricity — that is, the same electric current — decomposes chemically equivalent quantities of all the bodies which it traverses; hence the weights of elements separated in these electrolytes are to each other as their chemical equivalents."
1834	Heinrich Lenz	Applied an extension of the law of conservation of energy to the non-conservative forces in electromagnetic induction to give the direction of the induced electromotive force (emf) and current resulting from electromagnetic induction. The law provides a physical interpretation of the choice of sign in Faraday's law of induction (1831), indicating that the induced emf and the change in flux have opposite signs.
1834	Jean-Charles Peltier	Discovered what is now called the Peltier effect: the heating effect of an electric current at the junction of two different metals.
1838	Michael Faraday	Using Volta's battery, Farraday discovered "cathode rays" when, during an experiment, he passed current through a rarefied air filled glass tube and noticed a strange light arc starting at the anode (positive electrode) and ending at the cathode (negative electrode).
1839	Alexandre Edmond Becquerel	Observed the photoelectric effect via an electrode in a conductive solution exposed to light.
1852	Edward Frankland	Initiated the theory of valency by proposing that each element has a specific "combining power", e.g. some elements such as nitrogen tend to combine with three other elements (e.g. NO_3) while others may tend to combine with five (e.g. PO_5), and that each element strives to fulfill its combining power (valency) quota.
1857	Heinrich Geissler	Invented the Geissler tube.
1858	Julius Plücker	Published the first of his classical researches on the action of magnets on the electric discharge of rarefied gases in Geissler tubes. He found that the discharge caused a fluorescent glow to form on the glass walls of the vacuum tube, and that the glow could be made to shift by applying a magnetic field to the tube. It was later shown by Johann Wilhelm Hittorf that the glow was produced by rays emitted from one of the electrodes (the cathode).
1859	Gustav Kirchhoff	Stated the "black body problem", i.e. how does the intensity of the electromagnetic radiation emitted by a black body depend on the frequency of the radiation and the temperature of the body?
1865	Johann Josef Loschmidt	Estimated the average diameter of the molecules in air by a method that is equivalent to calculating the number of particles in a given volume of gas.[1] This latter value, the number density of particles in an ideal gas, is now called the Loschmidt constant in his honour, and is approximately proportional to the Avogadro constant. The connection with Loschmidt is the root of the symbol L sometimes used for the Avogadro constant, and German language literature may refer to both constants by the same name, distinguished only by the units of measurement.[2]
1868	Norman Lockyer and Edward Frankland	On October 20 observed a yellow line in the solar spectrum, which he named the "D3 Fraunhofer line" because it was near the known D1 and D2 lines of sodium. He correctly concluded that it was caused by an element in the Sun unknown on Earth. Lockyer and Frankland named the element with the Greek word for the Sun, ἥλιος, "helios."
1869	Dmitri Mendeleev	Devises the Periodic Table of the Elements.
1869	Johann Wilhelm Hittorf	Studied discharge tubes with energy rays extending from a negative electrode, the cathode. These rays, which he discovered but were later called cathode rays by Eugen Goldstein, produced a fluorescence when they hit a tube's glass walls and, when interrupted by a solid object, cast a shadow.
1869	William Crookes	Invented the Crookes tube.
1873	Willoughby Smith	Discovered the photoelectric effect in metals not in solution (i.e., selenium).
1873	James Clerk Maxwell	Published his theory of electromagnetism in which light was determined to be an electromagnetic wave (field) that could be propagated in a vacuum.
1877	Ludwig Boltzmann	Suggested that the energy states of a physical system could be discrete.
1879	William Crookes	Showed that cathode rays (1838), unlike light rays, can be bent in a magnetic field.
1885	Johann Balmer	Discovered that the four visible lines of the hydrogen spectrum could be assigned integers in a series

1886	Henri Moissan	Isolated elemental fluorine after almost 74 years of effort by other chemists.
1886	Oliver Heaviside	Coined the term "inductance."
1886	Eugen Goldstein	Goldstein had undertaken his own investigations of discharge tubes and had named the light emissions studied by others "kathodenstrahlen", or cathode rays. In 1886, he discovered that discharge tubes with a perforated cathode also emit a glow at the cathode end. Goldstein concluded that in addition to the already-known cathode rays (later recognized as electrons) moving from the negatively charged cathode toward the positively charged anode, there is another ray that travels in the opposite direction. Because these latter rays passed through the holes, or channels, in the cathode, Goldstein called them "kanalstrahlen", or canal rays. He determined that canal rays are composed of positive ions whose identity depends on the residual gas inside the tube. It was another of Helmholtz's students, Wilhelm Wien, who later conducted extensive studies of canal rays, and in time this work would become part of the basis for mass spectrometry.
1887	Albert A. Michelson and Edward W. Morley	Conducted what is now called the "Michelson-Morley" experiment, in which they disproved the existence of a luminiferous aether and that the speed of light remained constant relative to all inertial frames of reference. The full significance of this discovery was not understood until Albert Einstein published his Theory of Special Relativity.
1887	Heinrich Hertz	Discovered the production and reception of electromagnetic (EM) radio waves. His receiver consisted of a coil with a spark gap, where a spark would be seen upon detection of EM waves transmitted from another spark gap source.
1888	Johannes Rydberg	Modified the Balmer formula to include the other series of lines, producing the Rydberg formula
1891	Alfred Werner	Proposed a theory of affinity and valence in which affinity is an attractive force issuing from the center of the atom which acts uniformly from there towards all parts of the spherical surface of the central atom.
1892	Heinrich Hertz	Showed that cathode rays (1838) could pass through thin sheets of gold foil and produce appreciable luminosity on glass behind them.
1893	Alfred Werner	Showed that the number of atoms or groups associated with a central atom (the "coordination number") is often 4 or 6; other coordination numbers up to a maximum of 8 were known, but less frequent.
1893	Victor Schumann	Discovered the vacuum ultraviolet spectrum.
1895	Sir William Ramsay	Isolated helium on Earth by treating the mineral cleveite (a variety of uraninite with at least 10% rare earth elements) with mineral acids.
1895	Wilhelm Röntgen	Discovered X-rays with the use of a Crookes tube.
1896	Henri Becquerel	Discovered "radioactivity" a process in which, due to nuclear disintegration, certain elements or isotopes spontaneously emit one of three types of energetic entities: alpha particles (positive charge), beta particles (negative charge), and gamma particles (neutral charge).
1897	J. J. Thomson	Showed that cathode rays (1838) bend under the influence of both an electric field and a magnetic field. To explain this he suggested that cathode rays are negatively charged subatomic electrical particles or "corpuscles" (electrons), stripped from the atom; and in 1904 proposed the "plum pudding model" in which atoms have a positively charged amorphous mass (pudding) as a body embedded with negatively charged electrons (raisins) scattered throughout in the form of non-random rotating rings. Thomson also calculated the mass-to-charge ratio of the electron, paving the way for the precise determination of its electrical charge by Robert Andrews Millikan (1913).
1900	Max Planck	To explain black-body radiation (1862), he suggested that electromagnetic energy could only be emitted in quantized form, i.e. the energy could only be a multiple of an elementary unit $E = h\nu$, where h is Planck's constant and ν is the frequency of the radiation.
1901	Frederick Soddy and Ernest Rutherford	Discovered nuclear transmutation when they found that radioactive thorium was converting itself into radium through a process of nuclear decay.
1902	Gilbert N. Lewis	To explain the octet rule (1893), he developed the "cubical atom" theory in which electrons in the form of dots were positioned at the corner of a cube and suggested that single, double, or triple "bonds" result when two atoms are held together by multiple pairs of electrons (one pair for each bond) located between the two atoms (1916).
1904	J. J. Thomson	Articulated the "plumb-pudding" model of the atom that was later experimentally disproved by Rutherford (1907)
1904	Richard Abegg	Noted the pattern that the numerical difference between the maximum positive valence, such as +6 for H_2SO_4, and the maximum negative valence, such as -2 for H_2S, of an element tends to be eight (Abegg's rule).
1905	Albert Einstein	Determined the equivalence of matter and energy.
1905	Albert Einstein	First to explain the effects of Brownian motion as caused by the kinetic energy (i.e., movement) of atoms, which was subsequently, experimentally verified by Jean Baptiste Perrin, thereby settling the century-long dispute about the validity of John Dalton's atomic theory.
1905	Albert Einstein	Published his Special Theory of Relativity
1905	Albert Einstein	Explained the photoelectric effect (1839), i.e. that shining light on certain materials can function to eject electrons from the material, he postulated as based on Planck's quantum hypothesis (1900), that light itself consists of individual quantum particles (photons).
1907	Ernest Rutherford	To test the plum pudding model (1904), he fired positively charged alpha particles at gold foil and noticed that some bounced back, thus showing that atoms have a small-sized positively charged atomic nucleus at its center.

1907	Ernest Rutherford	To test the plum pudding model (1904), he fired positively charged alpha particles at gold foil and noticed that some bounced back, thus showing that atoms have a small-sized positively charged atomic nucleus at its center.
1909	Geoffrey Ingram Taylor	Demonstrated that interference patterns of light were generated even when the light energy introduced consisted of only one photon. This discovery of the wave–particle duality of matter and energy was fundamental to the later development of quantum field theory.
1909 and 1916	Albert Einstein	Showed that, if Planck's law of black-body radiation is accepted, the energy quanta must also carry momentum $p = h / \lambda$, making them full-fledged particles, albeit with no "rest mass."
1911	Lise Meitner and Otto Hahn	Performed an experiment that showed that the energies of electrons emitted by beta decay had a continuous rather than discrete spectrum. This was in apparent contradiction to the law of conservation of energy, as it appeared that energy was lost in the beta decay process. A second problem was that the spin of the Nitrogen-14 atom was 1, in contradiction to the Rutherford prediction of ½. These anomalies were later explained by the discoveries of the neutrino and the neutron.
1912	Henri Poincaré	Published an influential mathematical argument in support of the essential nature of energy quanta.[3][4]
1913	Robert Andrews Millikan	Publishes the results of his "oil drop" experiment, in which he precisely determines the electric charge of the electron. Determination of the fundamental unit of electric charge made it possible to calculate the Avogadro constant (which is the number of atoms or molecules in one mole of any substance) and thereby to determine the atomic weight of the atoms of each element.
1913	Niels Bohr	To explain the Rydberg formula (1888), which correctly modeled the light emission spectra of atomic hydrogen, Bohr hypothesized that negatively charged electrons revolve around a positively charged nucleus at certain fixed "quantum" distances and that each of these "spherical orbits" has a specific energy associated with it such that electron movements between orbits requires "quantum" emissions or absorptions of energy.
1911	Ştefan Procopiu	Performed experiments in which he determined the correct value of electron's magnetic dipole moment, $\mu_B = 9.27 \times 10^{-21}$ erg·Oe^{-1}
1916	Gilbert N. Lewis	Developed the Lewis dot structures that ultimately led to a complete understanding of the electronic covalent bond that forms the fundamental basis for our understanding of chemistry at the atomic level; he also coined the term "photon" in 1926.
1916	Arnold Sommerfeld	To account for the Zeeman effect (1896), i.e. that atomic absorption or emission spectral lines change when the light is first shone through a magnetic field, he suggested there might be "elliptical orbits" in atoms in addition to spherical orbits.
1918	Ernest Rutherford	Noticed that, when alpha particles were shot into nitrogen gas, his scintillation detectors showed the signatures of hydrogen nuclei. Rutherford determined that the only place this hydrogen could have come from was the nitrogen, and therefore nitrogen must contain hydrogen nuclei. He thus suggested that the hydrogen nucleus, which was known to have an atomic number of 1, was an elementary particle, which he decided must be the protons hypothesized by Eugen Goldstein (1886).
1919	Irving Langmuir	Building on the work of Lewis (1916), he coined the term "covalence" and postulated that coordinate covalent bonds occur when the electrons of a pair come from the same atom, thus explaining the fundamental nature of chemical bonding and molecular chemistry.
1922	Arthur Compton	Found that X-ray wavelengths increase due to scattering of the radiant energy by "free electrons." The scattered quanta have less energy than the quanta of the original ray. This discovery, now known as the "Compton effect" or "Compton scattering", demonstrates the "particle" concept of electromagnetic radiation.
1922	Otto Stern and Walther Gerlach	Stern–Gerlach experiment detects discrete values of angular momentum for atoms in the ground state passing through an inhomogeneous magnetic field leading to the discovery of the spin of the electron.
1923	Louis de Broglie	Postulated that electrons in motion are associated with waves the lengths of which are given by Planck's constant h divided by the momentum of the $mv = p$ of the electron: $\lambda = h / mv = h / p$.
1924	Satyendra Nath Bose	His work on quantum mechanics provides the foundation for Bose–Einstein statistics, the theory of the Bose–Einstein condensate, and the discovery of the boson.
1925	Friedrich Hund	Outlined the "rule of maximum multiplicity" which states that, when electrons are added successively to an atom, as many levels or orbits are singly occupied as possible before any pairing of electrons with opposite spin occurs, and also made the distinction that the inner electrons in molecules remain in their atomic orbitals and only the valence electrons need occupy the molecular orbitals involving both nuclei of the atoms participating in a covalent bond.
1925	Werner Heisenberg	Developed the matrix mechanics formulation of quantum mechanics.
1925	Wolfgang Pauli	Outlined the "Pauli exclusion principle" which states that no two identical fermions may occupy the same quantum state simultaneously.
1926	Gilbert Lewis	Coined the term photon, which he derived from the Greek word for light, φως (transliterated phôs).
1926	Erwin Schrödinger	Used De Broglie's electron wave postulate (1924) to develop a "wave equation" that represents mathematically the distribution of a charge of an electron distributed through space, being spherically symmetric or prominent in certain directions, i.e. directed valence bonds, which gave the correct values for spectral lines of the hydrogen atom.

1927	Charles Drummond Ellis (along with James Chadwick and colleagues)	Finally established clearly that the beta decay spectrum is really continuous, ending all controversies.
1927	Walter Heitler	Used Schrödinger's wave equation (1926) to show how two hydrogen atom wavefunctions join together, with plus, minus, and exchange terms, to form a covalent bond.
1927	Robert Mulliken	In 1927 Mulliken worked, in coordination with Hund, to develop a molecular orbital theory where electrons are assigned to states that extend over an entire molecule and in 1932 introduced many new molecular orbital terminologies, such as σ bond, π bond, and δ bond.
1928	Paul Dirac	In the Dirac equations, Paul Dirac integrated the principle of special relativity with quantum electrodynamics and thereby hypothesized the existence of the positron.
1928	Linus Pauling	Outlined the nature of the chemical bond in which he used Heitler's quantum mechanical covalent bond model (1927) to describe the quantum mechanical basis for all types of molecular structure and bonding, thereby suggesting that different types of bonds in molecules can become equalized by the rapid shifting of electrons, a process called "resonance" (1931), such that resonance hybrids contain contributions from the different possible electronic configurations.
1929	John Lennard-Jones	Introduced the linear combination of atomic orbitals approximation for the calculation of molecular orbitals.
1930	Wolfgang Pauli	In a famous letter written by him, Pauli suggested that, in addition to electrons and protons, atoms also contained an extremely light neutral particle which he called the "neutron". He suggested that this "neutron" was also emitted during beta decay and had simply not yet been observed. Later it was determined that this particle was actually the almost massless neutrino.
1931	Walther Bothe and Herbert Becker	Found that, if the very energetic alpha particles emitted from polonium fell on certain light elements, specifically beryllium, boron, or lithium, an unusually penetrating radiation was produced. At first this radiation was thought to be gamma radiation, although it was more penetrating than any gamma rays known, and the details of experimental results were very difficult to interpret on this basis. Some scientists began to hypothesize the possible existence of another fundamental, atomic particle.
1932	Irène Joliot-Curie and Frédéric Joliot	Showed that if the unknown radiation generated by alpha particles fell on paraffin or any other hydrogen-containing compound, it ejected protons of very high energy. This was not in itself inconsistent with the proposed gamma ray nature of the new radiation, but detailed quantitative analysis of the data became increasingly difficult to reconcile with such a hypothesis.
1932	Irène Joliot-Curie and Frédéric Joliot	Showed that if the unknown radiation generated by alpha particles fell on paraffin or any other hydrogen-containing compound, it ejected protons of very high energy. This was not in itself inconsistent with the proposed gamma ray nature of the new radiation, but detailed quantitative analysis of the data became increasingly difficult to reconcile with such a hypothesis.
1932	James Chadwick	Performed a series of experiments showing that the gamma ray hypothesis for the unknown radiation produced by alpha particles was untenable, and that the new particles must be the neutrons hypothesized by Enrico Fermi. Chadwick suggested that, in fact, the new radiation consisted of uncharged particles of approximately the same mass as the proton, and he performed a series of experiments verifying his suggestion.
1932	Werner Heisenberg	Applied perturbation theory to the two-electron problem and showed how resonance arising from electron exchange could explain exchange forces.
1932	Mark Oliphant	Building upon the nuclear transmutation experiments of Ernest Rutherford done a few years earlier, fusion of light nuclei (hydrogen isotopes) was first observed by Oliphant in 1932. The steps of the main cycle of nuclear fusion in stars were subsequently worked out by Hans Bethe throughout the remainder of that decade.
1932	Carl D. Anderson	Experimentally proves the existence of the positron.
1933	Leó Szilárd	First theorized the concept of a nuclear chain reaction. He filed a patent for his idea of a simple nuclear reactor the following year.
1934	Enrico Fermi	Studies the effects of bombarding uranium isotopes with neutrons.
1934	N. N. Semyonov	Develops the total quantitative chain chemical reaction theory. The idea of the chain reaction, developed by Semyonov, is the basis of various high technologies using the incineration of gas mixtures. The idea was also used for the description of the nuclear reaction.
1935	Hideki Yukawa	Published his hypothesis of the Yukawa Potential and predicted the existence of the pion, stating that such a potential arises from the exchange of a massive scalar field, such as would be found in the field of the pion. Prior to Yukawa's paper, it was believed that the scalar fields of the fundamental forces necessitated massless particles.
1936	Carl D. Anderson	Discovered muons while studying cosmic radiation.
1937	Carl Anderson	Experimentally proved the existence of the pion.
1938	Charles Coulson	Made the first accurate calculation of a molecular orbital wavefunction with the hydrogen molecule.

1938	Otto Hahn, Fritz Strassmann, Lise Meitner, and Otto Robert Frisch	Hahn and Strassmann sent a manuscript to Naturwissenschaften reporting they had detected the element barium after bombarding uranium with neutrons. Simultaneously, they communicated these results to Meitner. Meitner, and her nephew Frisch, correctly interpreted these results as being nuclear fission. Frisch confirmed this experimentally on 13 January 1939.
1939	Leó Szilárd and Enrico Fermi	Discovered neutron multiplication in uranium, proving that a chain reaction was indeed possible.
1942	Kan-Chang Wang	First proposed the use of beta capture to experimentally detect neutrinos.
1942	Enrico Fermi	Created the first artificial self-sustaining nuclear chain reaction, called Chicago Pile-1 (CP-1), in a racquets court below the bleachers of Stagg Field at the University of Chicago on December 2, 1942.
1945	Manhattan Project	First nuclear fission explosion.
1947	G. D. Rochester and C. C. Butler	Published two cloud chamber photographs of cosmic ray-induced events, one showing what appeared to be a neutral particle decaying into two charged pions, and one which appeared to be a charged particle decaying into a charged pion and something neutral. The estimated mass of the new particles was very rough, about half a proton's mass. More examples of these "V-particles" were slow in coming, and they were soon given the name kaons.
1948	Sin-Itiro Tomonaga and Julian Schwinger	Independently introduced perturbative renormalization as a method of correcting the original Lagrangian of a quantum field theory so as to eliminate an infinite series of counterterms that would otherwise result.
1951	Clemens C. J. Roothaan and George G. Hall	Derived the Roothaan-Hall equations, putting rigorous molecular orbital methods on a firm basis.
1952	Manhattan Project	First explosion of a thermonuclear bomb.
1952	Herbert S. Gutowsky	Physical chemistry of solids investigated by NMR: structure, spectroscopy and relaxation
1952	Charles P. Slichter	Introduced Chemical shifts, NQR in solids, the first NOE experiments
1952	Albert W. Overhauser	First investigation of dynamic polarization in solids/NOE-Nuclear Overhauser Effect
1953	Charles H. Townes, (collaborating with J. P. Gordon, and H. J. Zeiger)	Built and reported the first ammonia maser; received a Nobel prize in 1964 for his experimental success in producing coherent radiation by atoms and molecules.
*1958—1959	Edward Raymond Andrew, A. Bradbury, and R. G. Eades; and independently, I. J. Lowe	described the technique of magic angle spinning.[5]
1956	P. Kuroda	Predicted that self-sustaining nuclear chain reactions should occur in natural uranium deposits.
1956	Clyde L. Cowan and Frederick Reines	Experimentally proved the existence of the neutrino.
1957	William Alfred Fowler, Margaret Burbidge, Geoffrey Burbidge, and Fred Hoyle	In their 1957 paper Synthesis of the Elements in Stars, they explained how the abundances of essentially all but the lightest chemical elements could be explained by the process of nucleosynthesis in stars.
1961	Clauss Jönsson	Performed Young's double-slit experiment (1909) for the first time with particles other than photons by using electrons and with similar results, confirming that massive particles also behaved according to the wave–particle duality that is a fundamental principle of quantum field theory.
1964	Murray Gell-Mann and George Zweig	Independently proposed the quark model of hadrons, predicting the arbitrarily named up, down, and strange quarks. Gell-Mann is credited with coining the term "quark", which he found in James Joyce's book Finnegans Wake.

1968	Stanford University	Deep inelastic scattering experiments at the Stanford Linear Accelerator Center (SLAC) showed that the proton contained much smaller, point-like objects and was therefore not an elementary particle. Physicists at the time were reluctant to identify these objects with quarks, instead calling them "partons" — a term coined by Richard Feynman. The objects that were observed at SLAC would later be identified as up and down quarks. Nevertheless, "parton" remains in use as a collective term for the constituents of hadrons (quarks, antiquarks, and gluons). The strange quark's existence was indirectly validated by the SLAC's scattering experiments: not only was it a necessary component of Gell-Mann and Zweig's three-quark model, but it provided an explanation for the kaon (K) and pion (π) hadrons discovered in cosmic rays in 1947.
1974	Pier Giorgio Merli	Performed Young's double-slit experiment (1909) using a single electron with similar results, confirming the existence of quantum fields for massive particles.
1995	Eric Cornell, Carl Wieman and Wolfgang Ketterle	The first "pure" Bose–Einstein condensate was created by Eric Cornell, Carl Wieman, and co-workers at JILA. They did this by cooling a dilute vapor consisting of approximately two thousand rubidium-87 atoms to below 170 nK using a combination of laser cooling and magnetic evaporative cooling. About four months later, an independent effort led by Wolfgang Ketterle at MIT created a condensate made of sodium-23. Ketterle's condensate had about a hundred times more atoms, allowing him to obtain several important results such as the observation of quantum mechanical interference between two different condensates.
2000	CERN	CERN scientists published experimental results in which they claimed to have observed indirect evidence of the existence of a quark–gluon plasma, which they call a "new state of matter".

Chapter 23

Timeline of quantum computing

This is a **timeline of quantum computing**.

23.1 1970s

- 1970

    - Stephen Wiesner invents conjugate coding.

- 1973

    - Alexander Holevo publishes a paper showing that n qubits cannot carry more than n classical bits of information (a result known as "Holevo's theorem" or "Holevo's bound").

    - Charles H. Bennett shows that computation can be done reversibly.

- 1975

    - R. P. Poplavskii publishes "Thermodynamical models of information processing" (in Russian)[1] which showed the computational infeasibility of simulating quantum systems on classical computers, due to the superposition principle.

- 1976

    - Polish mathematical physicist Roman Stanisław Ingarden publishes a seminal paper entitled "Quantum Information Theory" in Reports on Mathematical Physics, vol. 10, 43–72, 1976. (The paper was submitted in 1975.) It is one of the first attempts at creating a quantum information theory, showing that Shannon information theory cannot directly be generalized to the quantum case, but rather that it is possible to construct a quantum information theory, which is a generalization of Shannon's theory, within the formalism of a generalized quantum mechanics of open systems and a generalized concept of observables (the so-called semi-observables).

23.2 1980s

- 1980

    - Paul Benioff described quantum mechanical Hamiltonian models of computers [2]
    - Yuri Manin proposed an idea of quantum computing[3]

- 1981

www.ingramcontent.com/pod-product-compliance
Lightning Source LLC
Chambersburg PA
CBHW080658190526

45169CB00006B/2168